Integration of ICT in Smart Organizations

István Mezgár
Hungarian Academy of Sciences and
Budapest University of Technology, Hungary

IDEA GROUP PUBLISHING
Hershey • London • Melbourne • Singapore

Acquisitions Editor: Michelle Potter
Development Editor: Kristin Roth
Senior Managing Editor: Amanda Appicello
Managing Editor: Jennifer Neidig
Copy Editor: Killian Piraro
Typesetter: Diane Huskinson
Cover Design: Lisa Tosheff
Printed at: Yurchak Printing Inc.

Published in the United States of America by
Idea Group Publishing (an imprint of Idea Group Inc.)
701 E. Chocolate Avenue
Hershey PA 17033
Tel: 717-533-8845
Fax: 717-533-8661
E-mail: cust@idea-group.com
Web site: http://www.idea-group.com

and in the United Kingdom by
Idea Group Publishing (an imprint of Idea Group Inc.)
3 Henrietta Street
Covent Garden
London WC2E 8LU
Tel: 44 20 7240 0856
Fax: 44 20 7379 0609
Web site: http://www.eurospanonline.com

Library of Congress Cataloging-in-Publication Data

Integration of ICT in smart organizations / Istvan Mezgar, editor.
p. cm.
Summary: "The book covers the state-of-the-art concepts and methodologies of smart organization development featuring information and communication technologies"--Provided by publisher.
Includes bibliographical references and index.
ISBN 1-59140-390-1 (hardcover) -- ISBN 1-59140-391-X (softcover) -- ISBN 1-59140-392-8 (ebook)
1. Information technology--Management. 2. Communication in organizations--Technological innovations. 3. Communication in management. 4. Management information systems. 5. Information networks--Management. 6. Knowledge management. I. Mezgár, István, 1951-
HD30.2.I547 2006
658.4'038--dc22

2005027414

British Cataloguing in Publication Data
A Cataloguing in Publication record for this book is available from the British Library.

All work contributed to this book is new, previously-unpublished material. The views expressed in this book are those of the authors, but not necessarily of the publisher.

Integration of ICT in Smart Organizations

Table of Contents

Preface vi

Section I: Main Characteristics of Smart Organizations

Chapter I
Smart Organizations in the Digital Age 1
Erastos Filos, Directorate-General Information Society and Media, European Commission, Belgium

Section II: Technologies for Operation of Smart Organizations

Chapter II
Applications of Agent-Based Technologies in Smart Organizations 39
László Zsolt Varga, Hungarian Academy of Sciences, Hungary

Chapter III
The HUB as an Enabling IT Strategy to Achieve Smart Organizations 68
Arturo Molina, ITESM, Campus Monterrey, Mexico
Ricardo Mejía, ITESM, Campus Monterrey, Mexico
Nathalíe Galeano, ITESM, Campus Monterrey, Mexico
Teresa Nájera, ITESM, Campus Monterrey, Mexico
Marcela Velandia, DaimlerChrysler Vehículos Comerciales México S.A de C.V., Mexico

Section III: Knowledge- and Human-Centered Technologies in Smart Organizations

Chapter IV
Knowledge Management in Smart Organizations **101**
Shirley Chan, The University of Hong Kong, Hong Kong

Chapter V
Bridging Diversity across Time and Space: The Case of Multidisciplinary Virtual Teams **136**
Violina Ratcheva, The University of Sheffield, UK

Chapter VI
Neural Data Mining System for Trust-Based Evaluation in Smart Organizations **159**
T. T. Wong, The Hong Kong Polytechnic University, Hong Kong

Section IV: Communication and Security Technologies for Smart Organizations

Chapter VII
New Challenges for Smart Organizations: Demands for Mobility – Wireless Communication Technologies **187**
István Mezgár, Hungarian Academy of Sciences and Budapest University of Technology and Economics, Hungary

Chapter VIII
Infrastructure Support for Smart Organizations: Integration of Web Service Partners in Heterogeneous Environments **257**
Peter Bertok, Royal Melbourne Institute of Technology, Australia
Xinjian Xu, Royal Melbourne Institute of Technology, Australia

Chapter IX
Grid Technology for Smart Organizations **289**
Gergely Sipos, Hungarian Academy of Sciences, Hungary
Péter Kacsuk, Hungarian Academy of Sciences, Hungary

Chapter X
Communication Security Technologies in Smart Organizations 333
Raphael C. W. Phan, Swinburne University of Technology (Sarawak Campus), Malaysia

About the Authors 354

Index 360

Preface

The production and service of the 21st century is based on distributed or networked organizations. The denomination of these organizations can be different (extended-, virtual-, smart-organization, etc.), but there are some main important common characteristics. In this type of organizations, flexible, independent organizational and production units are working together, reacting in an intelligent way to the challenges and uncertainty of the environment while using some type of communication network (wired or wireless). In case of the so-called "smart organizations," the integration of information and communication technologies (ICT), knowledge and organizational networks form the background of operation.

As the base of networked organizations is the interdependent, separate production and service teams and units, the cooperation and collaboration among them is of vital importance. The structure, the communication systems, and the collaborating people, teams, and organizations that define today's organizations characteristics must be harmonized to accomplish complex, demanding tasks. The collaboration means contacts among users, so human beings have outstanding importance in the operation.

According to experience, the improper application of this human factor can make the operation very inefficient, even in the case of the technically most advanced systems. The lowest level of connection among systems is made through protocols; the highest contact level is among the decision-makers, the human connections. A very important element of this human contact is the trust. In a smart organization, trust is the flavor, the medium in which players are moving. Only trust can bridge the cultural, geographical, and organizational distances of humans (and even of firms) avoiding problematic situations.

Due to the rapidly developing information and communication technologies, the complexity of the networked organizations are becoming very high, so the representation of their structure, the description of their operation, and their control needs new technologies and approaches. In today's turbulent environment, only those organizations that effectively apply the results of the different disciplines can survive. Smart organization belongs to this kind of category.

The purpose of the book is to introduce the methodologies, approaches for describing the structure, the smooth operation, the communication, and knowledge-handling of smart organizations in an integrated way. The value of the book is bringing together the theories applied in different fields. In the book, the main regulation aspect was how these theories can be applied in describing and solving various demands of smart organization. The book covers the state-of-the-art concepts and methodologies of smart organization development, taking into account the current results in information and communication technologies, and will outlinc also the trends of the fields.

Organization of the Book

The book contains ten chapters written by professional researchers coming from the field of academics and industry. The chapters have been organized into four interrelated sections.

Section I: Main Characteristics of Smart Organizations. The chapter in this section makes an overview of the main characteristics of smart organizations.

Chapter I. The chapter titled "Smart Organizations in the Digital Age" authored by Filos presents and explains the concept of the smart organization. This concept arose from the need for organizations to respond dynamically to the changing landscape of a digital economy. A smart organization is understood to be both internetworked and knowledge-driven, and therefore able to adapt to new organizational challenges rapidly and sufficiently to create and exploit knowledge in response to opportunities of the digital age. The three networking dimensions of smart organizations, ICT-enabled virtuality, organizational teaming, and knowledge hyperlinking, are elaborated. This networking capability allows smart organizations to cope with complexity and with rapidly changing economic environments. The chapter also shows how managing the smart organization requires a more "fuzzy" approach to managing smart resources: people, information, knowledge, and creativity. Some research work is also presented, mainly from the European perspective.

Section II: Technologies for Operation of Smart Organizations. These chapters introduce technologies appropriate to increasing the effectiveness of the operation of smart organizations.

Chapter II. The dynamism is an important factor of smart organizations. Varga, in his chapter "Applications of Agent-Based Technologies in Smart Organizations," introduces agent technology as a means of creating dynamic software systems for the changing needs of smart organizations. The notion of agency is introduced, and individual and collective agent architectures are described. Agent interaction methods and agent system design techniques are discussed. Application areas of agent technology are overviewed. The chapter argues that the autonomous and proactive nature of agent systems makes them suitable as the new information infrastructure for the networked components of dynamically changing smart organizations.

Chapter III. In this chapter, "The HUB as an Enabling IT Strategy to Achieve Smart Organizations," Molina, Mejía, Galeano, Najera, and Velandia introduce the concept of Virtual Enterprise Broker (VEB) supported by the use of a "hub" of integrated e-services as an enabling IT strategy to design and create smart organizations. The VEB model is described in terms of core processes, success measures, and supporting information and communication technologies. The VEB is a business entity that enables the design, configuration, creation, and operation of smart organizations. VEB core processes are supported by e-services integrated in a "hub" (the concept of hub refers to a proposed centre of integrated e-services for virtual business) that is supported by Web-based applications and technologies. Six integrated e-services have been defined, based on the concept of on-demand services for value added industrial networks: e-marketing, e-brokerage, e-planning, e-engineering, e-supply and e-productivity. The conjunction of these e-services improves industrial networks performance. A description of the e-services and hub architecture is presented in detail.

Section III: Knowledge- and Human-Centered Technologies in Smart Organizations. Two chapters in this section deal with different aspects of handling knowledge in smart organizations. The third chapter focuses on the role of trust in smart organization.

Chapter IV. Handling and management of knowledge is a basic task in smart organizations. The chapter "Knowledge Management in Smart Organizations," authored by Chan, looks at the deployment of appropriate information and communication technologies in helping smart organizations manage knowledge. Taking a management perspective, smart organizations can be regarded as those that can make smart strategic decisions and put into practice such managerial principles as value creation, continual learning, embracing uncertainty, and empowerment. Making good decisions would involve gathering and synthesizing the appropriate knowledge—knowledge about the market, products, suppliers, customers, competitors, and others. Different schools of knowledge-management theories and the related technologies are discussed.

Chapter V. Virtual teams are basic units of networked organizations. The uniqueness of multidisciplinary teamwork is in its potential to integrate different bodies of knowledge into a new synergy. However, previous empirical studies have shown that member heterogeneity and geographic separation hinder effective sharing and use of team knowledge. In "Bridging Diversity across Time and Space: The Case of Multidisciplinary Virtual Teams," Ratcheva explores how such teams interact to overcome the barriers and take advantage of their "built-in" knowledge diversity. The findings indicate that often teams lack common background knowledge at the beginning of the projects, and in order to resolve differences members rely on their external intellectual and social communities. The reported research establishes a positive correlation between team members' participation in multiple professional and social networks, and teams' abilities to successfully build on their knowledge diversity. The findings also suggest a need to reconceptualize the boundaries of multidisciplinary teams and to consider the processes of sharing diverse knowledge in a wider social context.

Chapter VI. Nowadays, many enterprises manufacture and distribute their products or services globally, and quite a number of smart organizations are formed on the Internet and are expected to evolve to a strategically important e-business model. Although information and communication technologies and knowledge management play an important role in linking the core and partner companies, it remains subservient to the humans that form the smart organizations. The "Neural Data Mining System for Trust-Based Evaluation in Smart Organizations" chapter, authored by Wong, identifies two instances in which trust-based evaluations of partners in the smart organizations are applicable. A review of the literature indicates that neither researchers nor practitioners agree on a single model of interfirm trust that applies to all partner evaluation contexts. A decision-support system based on neural network and data mining technologies is proposed. A case example is given to illustrate a trust-based evaluation in a real situation.

Section IV: Communication and Security Technologies for Smart Organizations. One of the three basic factors of smart organization, the communication technologies and their security, is introduced in this section.

Chapter VII. The chapter entitled "New Challenges for Smart Organizations: Demands for Mobility – Wireless Communication Technologies," written by Mezgár, introduces the different types of wireless technologies that can be applied in smart organizations (SO). Smart organization is an outstanding representative of networked organizations, as its organization structure, communication, and knowledge-based applications are coordinated and all networked. The chapter describes the communication demands of SO, taking care on wired and especially wireless networks that offer mobility for users. Access at any time from anywhere to enterprise information for registered users guarantees mobility, a basic demand for a dynamic organization today. Security, trust, and interoperability aspects are also discussed as important characteristics of the up-to-date infocom systems. Finally, the main impacts of wireless technologies on smart organizations are summarized. Through the survey of structure and operation of wireless technologies and their impacts, it is easy to understand that wireless communication technology has a strategic role in the effective, competitive operation of networked organizations.

Chapter VIII. In a rapidly changing world, continuous adoption of new practices is crucial for survival; organizations embracing the latest technologies have a competitive edge. Smart organizations readily take onboard new organizational forms and practices, those in particular that offer agility and responsiveness. The Internet and the World Wide Web offer a new way of collaboration via Web services, but heterogeneity of different service components make cooperation difficult. Bertok and Xu describe in this chapter "Infrastructure Support for Smart Organizations: Integration of Web Service Partners in Heterogeneous Environments," a new approach to combine Web services by employing a layered structure in which composition of a value-added service can be built from individual components, and each service component can have semantically equivalent but syntactically different alternatives.

Chapter IX. In the past few years, grid computing and grid development have become one of the most remarkable and most generously financed topics within computer science. At the same time, only the most well-informed IT experts and researchers know what it really means and tries to achieve. In the "Grid Technology for Smart Organizations" chapter, Sipos and Kacsuk make a difference. First, the chapter discusses the basic goal of grid computing, then shows the latest, service-oriented grid approach by introducing two technologies that have been developed for distributed systems. The first one is Web services and its grid extensions OGSA, while the other one is Jini. In the second part of the chapter, the authors introduce their prediction about the future of grid computing and the basic role it will probably have in the life of smart organizations.

Chapter X. In this chapter, "Communication Security Technologies in Smart Organizations," Phan introduces the security technologies that are important in guaranteeing the high quality of communication within smart organizations. First, the various forms of communication that can be used in the current information age are briefly reviewed before outlining the possible threats that can be faced in each communication medium. Then, the relevant security technologies are described that help to protect communication media from common threats, as well as the security tools available in the market that implement these technologies. The topics discussed in this chapter would serve to educate the smart organizations toward securing their various means of communication, which is vital for a business establishment to exist and coexist with peers and partners.

The editor hopes that the book will be a useful summary of ideas and foresights needed to develop and operate smart organizations. In the book, there are detailed discussions of different methodologies, concepts, and technologies required for handling and exchanging knowledge and information, and for safe communication via different media in virtual environments and in smart organizations.

The chapters offer practical suggestions for developing and operating different subsystems of smart organizations. Thus, undergraduate and graduate students could use the book when taking courses in knowledge management, communication technologies, networked organizations, and some related areas. Practitioners also could be interested when seeking to better support and raise the level of their decision-making processes. Applying an existing theory in a new field or integrating different theories to solve a new problem always generates additional motivation. Hopefully, there will be also some results introduced that can generate new ideas in the readers, inspiring new research works or new directions as well.

István Mezgár, Editor
December 2005
Budapest, Hungary

Acknowledgments

First of all, the editor would like to thank the efforts and thorough work of all authors, their invested time and knowledge (I know that some of them have written parts of their manuscripts during their holidays), and the positive attitude to transfer their knowledge and expertise to other people, to the readers. Working with them on this project was an extraordinary experience.

A further special note of thanks goes also to all the staff at Idea Group Inc., whose contributions throughout the whole process from inception of the initial idea to final publication have been invaluable. At the beginning, it was Mehdi Khosrow-Pour who supported the idea to launch this project, and later on at critical phases Jan Travers gave me trust and support that filled me up with new energies to go on with the work.

The management and editing work of this book took me far more time and effort than I estimated when I started the project. I owe a great debt to Michele Rossi and Kristin Roth, the successive development editors of this book. They organized and carried out the complex tasks of editorial management and deadline coordination, and gave me useful, practical advice when it seemed some tasks were a dead end.

I would like to acknowledge the work of all involved in the review process of the book, without whose support the project could not have been satisfactorily completed. Most of the authors of chapters included in this book also served as referees for articles written by other authors. Thanks go to all those who provided constructive and comprehensive reviews.

Finally, I have to thank my colleagues and my environment for tolerating my book-editing "hobby" for so long.

István Mezgár, Editor

December 2005

Budapest, Hungary

Section I

Main Characteristics of Smart Organizations

Chapter I

Smart Organizations in the Digital Age

Erastos Filos, Directorate-General Information Society and Media, European Commission, Belgium

Abstract

The chapter aims to present and explain the concept of the smart organization. This concept arose from the need for organizations to respond dynamically to the changing landscape of a digital economy. A smart organization is understood to be both internetworked and knowledge-driven, and therefore able to adapt to new organizational challenges rapidly. It is sufficiently agile to respond to opportunities of the digital age. The three networking dimensions of smart organizations, ICT-enabled virtuality, organizational teaming, and knowledge hyperlinking, are elaborated. This networking capability allows smart organizations to cope with complexity and with rapidly changing economic environments. The paper also shows how managing the smart organization requires a more "fuzzy" approach to managing smart resources: people, information, knowledge, and creativity. Research is also presented, mainly from the European perspective. It has been key to creating the conditions for organizations to become smart.

Characteristics of the Digital Age

Over the last decades, information and communication technologies (ICT) have been the enabling factor in organizational change and innovation, and there is now evidence of their impact on industrial value chains. Organizations today strive to become agile and to operate profitably in an increasingly competitive environment of continuously and unpredictably changing markets.

The digital age is different from the industrial age in various ways (Figure 1). For example, today ICT represent a substantial—and increasing—part of the added value of products and services. ICT-intensive sectors include manufacturing, automotive, aerospace, pharmaceuticals, medical equipment, and agro-food, as well as financial services, media, and retail. In the automotive sector, for instance, an estimated 70% of innovations that happened over the last 20 years were related to ICT.

According to recent studies, more than half of the productivity gains in developed economies can be attributed to ICT (OECD, 2003; O'Mahony & van Ark, 2003). The gains stem both from the production of innovative, high-value goods and services based on ICT, as well as from improvements in business processes through a wider diffusion, adoption and use of ICT across the economy. Their impact on the economy and on society at large has led to remarkable changes.

Figure 1. Industrial vs. digital age characteristics

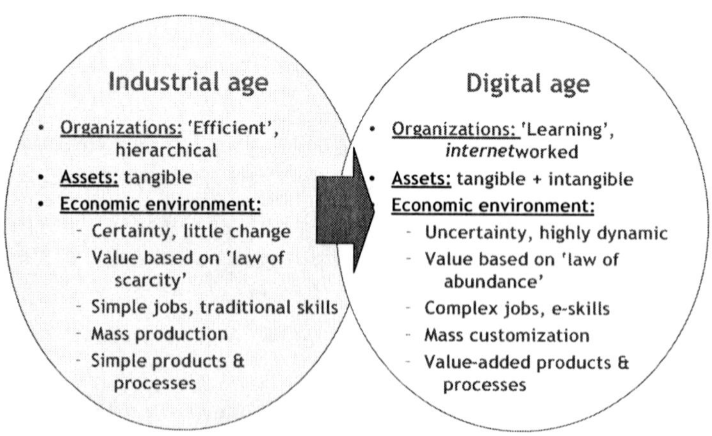

A "Hyperlinked" Economy

The increased networking in a global economy is due to the pervasiveness of ICT and the Internet. Since business success depends on the ability to innovate, and since innovation comes from a clash of ideas, networks provide a natural environment for this. The Internet not only facilitates a hyperlinking of documents, but also a hyperlinking of people and of organizations (Levine, Locke, Searls, & Weinberger, 2000). The *internet*worked economy (Ticoll, Lowy, & Kalacota, 1998) is about the right set of connections between people and organizations in whatever role they may be in. In relationships that are fostered via networks, roles become blurred: The seller becomes the "buyer" of valuable feedback on his product. Smart business organizations today see customers, suppliers, regulators, and even competitors as stakeholders who can make valuable contributions to their success.

"Value" Redefined

Individuals and organizations today understand value as something different from value in its traditional sense—that is, not only attributable to something that is unique or scarce. Value in a networked economy grows with the number of intermediation opportunities (e.g., relationships). Network theory predicts an exponential growth of interactions with a growing number of involved members ("nodes"). The more nodes there are in a network community, the more each node becomes an intermediary to all others (Kelly, 1999).

Another reason for the new perception of value is the fact that economic value is no longer derived from tangible assets alone—for example, from investments in labor, plants, and machinery. "Smart" resources—such as information, content, software, knowledge, brands, and innovation capability—contribute increasingly to value creation in today's economy.

Intangible Assets

Brands and knowledge are becoming a source of value, not unlike capital. Brands, for example, represent accumulated surplus value turned into client loyalty, which translates into lower marketing costs, higher prices, or larger market share for the owner organization (Davis & Meyer, 1998). In digital markets, brands are an invaluable source of trust and orientation to consumers who are looking for quality and security. Many organizations invest heavily in building a reputation that is conveyed through a brand. Some businesses have even outsourced almost all other activities just to maintain their focus on managing the brand. In an *internet*worked economy, knowledge is a key intangible asset that requires effort to develop and to protect.

The Growing Need for Trust

A key question in the digital economy is: "How can you do business with somebody that you do not see?" (Handy, 1995). As business relies more and more on technologies and infrastructures that reduce geographical distance, open communication networks and associated information systems become vulnerable to integrity and security threats. Technologically, trust and dependability must be established and maintained through security technologies such as cryptography and electronic authentication (biometrics, electronic signatures, etc.) and by technologies that enhance privacy and help protect and manage intellectual rights, digital assets, and identities. In the socio-organizational context, trust becomes an essential element of management.

The Smart Organization

Most organizations are not designed—they evolve. This is why biological analogies may provide an appropriate means to describe organization phenomena. But not all organizations adapt equally well to the environment within which they evolve. Many, like dinosaurs of great size but with little brains, remain unchanged in a changing world. In a digital economy, the law of survival of the fittest will evidence its relevance to organizations as it does in the biological domain.

Handy (1999) sees the old understanding of alliances with suppliers, consultants, retailers, and agents changing into a new type—that is, stakeholder alliances with suppliers, customers, and employees, as well as alliances with competitors. As no organization today can afford to remain an "island entire unto itself," every organization is a network of other organizations. No discussion of structure can therefore rest content with the inside of the organization.

Some organizational metaphors include terms like *adhocracy* (Mintzberg, 1980), *cluster organization* (Mills, 1991), *network organization* (Foy, 1980; Imai & Itami, 1984), and *organizational marketplace* (Williamson, 1975). All these concepts share certain common characteristics, like flatter hierarchies, dynamic structures, empowerment of individuals, and high esteem of individuals' capabilities, intellect, and knowledge. However, although they may gain importance in the digital age, they cannot be considered a panacea to cure all management ills.

Despite the proposed new models, the basic duality between a hierarchical (bureaucratic) and a networked structure remains. In *The Knowledge-Creating Company*, Nonaka and Takeuchi (1995) argue that while for most of the 20th century organizational structures have oscillated between these two basic types, what is necessary for knowledge-driven organizations today is a smart combination of both. They propose the concept of the hyperlinked organization, which is able to maximize corporate-level (hierarchical) efficiency and local flexibility (networked teams) as it grows in scale and complexity while maintaining its basic capability to create value.

The implications of the above trends for organizations have led to a proliferation of adjectives applied primarily to enterprises—among others, the agile enterprise, networked organization, virtual company, extended enterprise, ascendant organization (Wickens, 1998), knowledge enterprise (Nonaka & Takeuchi, 1995), learning organization (Senge, 1990), ambidextrous organization (O'Reilly & Tushman, 2004). The definitions all have their nuances, deriving from the emphasis on one or another combination of the aspects above. Ultimately, however, they all point to the need to respond to the changing landscape of the digital economy in dynamic and innovative ways.

Within the European Commission's research program Information Society Technologies (IST, 2002), the term "smart organization" was coined for organizations that are knowledge-driven, *internet*worked, and dynamically adaptive to new organizational forms and practices, learning as well as agile in their ability to create and exploit the opportunities offered in the digital age. Smart organizations involve more than the capability of setting up and exploiting a digital infrastructure or the ability to enter into a virtual collaboration with other partner organizations (Filos & Banahan, 2001b).

The Three Networking Dimensions of Smart Organizations

Smart organizations are networked in three dimensions: the ICT dimension, the organizational dimension, and the knowledge dimension (see Figure 2).

Networking at the ICT level enables organizations to move into extended or virtual organizational forms. This may not be enough, though, since the organizational structure and management cultures may need to move beyond steep hierarchies toward leaner business processes organized around flexible cross-functional teams. A further step lies in involving the knowledge dimension into the networking by empowering the individuals in those teams to become dynamically linked to each other and to share information and knowledge (Savage, 1996).

ICT-Enabled Virtuality

Smart organizations have the capability to enter into a virtual collaboration with other organizations. Virtual organizational forms are thus an essential characteristic of smart organizations in the digital age (Filos, 2005).

While in the past the aim for organizations was to integrate the supply chain as tightly as possible, the focus is now shifting from vertical integration toward *internet*worked organizational forms. One characteristic is a focus on "core business", while noncore activities are "outsourced" over the Internet and through e-business exchanges to partners that may have the capability to perform specific tasks better or more cost-effectively.

Figure 2. Smart organizations are networked in three dimensions (Filos & Banahan, 2001a)

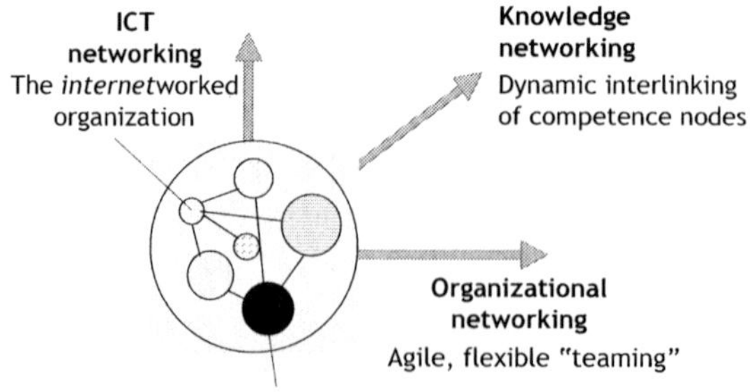

Organizational Teaming

For businesses, large and small, collaborative partnerships have become central to competitive success in fast-changing global markets. Since many of the skills and resources essential to an organization's competence lie outside its boundaries, and outside management's direct control, partnerships are no longer an option but a necessity. Organizations today have to be "smart" in their ability to conceive, shape, and sustain a wide variety of collaborative partnerships. Hence the challenge: The "capacity to collaborate" becomes a core competence of an organization.

Collaborative partnerships are held together because of the added value they offer. Organizations that enter into a cooperation with others do so because of a variety of strategic goals they may pursue (Doz & Hamel, 1998). These can be:

- Resource optimization (sharing investment with regard to infrastructure, R&D, market knowledge and the sharing of risks, while maintaining the focus on one's own core competences)
- Creation of synergies, by bundling complementary competences and by offering customers a solution rather than a mere product or a service
- Attaining critical mass in terms of capital investment, shared markets, and customers
- Achieving increased benefits in terms of shorter time-to-market, higher quality, with less investment

Goldman, Nagel, and Preiss (1995) have described four strategic dimensions of agile behavior that are crucial to smart organizations. These are customer focus, commitment to intra- and inter-organizational collaboration, organizing to master change and uncertainty, and leveraging the impact of people (entrepreneurial culture) and knowledge (intellectual capital).

Knowledge Hyperlinking

Nonaka and Takeuchi (1995) see as a basic precondition for the growth of organizational knowledge the creation of a "hyper-text" organization, which is made up of three interconnected layers or contexts, such as the business system, the project teams, and the (corporate) knowledge base. The key characteristic of the knowledge-creating company is this capability to shift contexts. The bureaucratic structure efficiently implements, exploits, and accumulates new knowledge through internalization and combination. Project teams generate (via externalization) conceptual and (via socialization) synthesized knowledge. The efficiency and stability of the bureaucracy is combined in this model with the effectiveness and dynamism of the project team. But, according to Nonaka and Takeuchi, these two elements are not sufficient without the third context, the knowledge base, which serves as a "clearing-house" for new knowledge to be generated inside both the enterprise and the project team contexts.

This hyperlinked organization has the organizational capability to convert knowledge from outside the organization by being an open system that features also continuous and dynamic knowledge interaction with partners outside the organization.

With the evolution of new organizational forms, such as networks, communities, and partnerships, the focus shifts from an ICT-centered to a human-centered perspective of knowledge management (KM). The knowledge sharing process is driven by people who work in a community that shares common interests and objectives. Evans and Roth (2004) elaborate on the basic premises and working principles of collaborative knowledge networks, which link communities together by providing a technical and social infrastructure for collaboration and knowledge management. Organizations that have implemented such environments report significant benefits in terms of knowledge transfer efficiency, response time, and innovation (Deloitte, 2002).

Lessons Learned from the Science of Complexity

The digital age is characterized by uncertainty and unpredictability, and organizations have to cope with it. This factor is radically changing the ways in which organizations relate to each other, and to the individuals who provide their core competence, and to their environment.

Sustainable innovation is the result of persistent disequilibrium between chaos and order. The *internet*worked economy resembles an ecology of organisms, interlinked and coevolving, constantly in flux, deeply tangled, ever expanding at its edges.

In their book *The Complexity Advantage*, Kelly and Allison (1999) discuss how six concepts derived from complexity science can be applied to business:

- In *nonlinear dynamics*, small differences at the start may lead to vastly different results. The so-called "butterfly effect" may prove valuable for business, particularly at turning points, such as the launch of a new product, the starting of a new division or investment in a new line of research.
- An *open system* is one in which the boundaries permit interaction with the environment. A good example for this is the living cell in a biological organism. Many organizations seem only partially open. Businesses, teams, leaders often shut out certain kinds of information and are open only to information that matches the way in which they see the world. However, it is critical for business organizations to also see the changing nature of their customers, markets, and competition in order to be able to offer genuine value.
- A *feedback loop* is simply a series of actions, each of which builds on the results of prior action and loops back in a circle to affect the original state. The final action either reinforces or changes the direction of the status quo. For example, although innovation is an important aspect of business success, an amplifying feedback loop might exaggerate the amount of innovation to the point at which nothing is ever produced or brought to the market. It is essential to identify such amplification and counterbalance it. Feedback loops, whether functional or dysfunctional, are a key part of the self-organization that emerges in all business.
- *Fractal structures* are those in which the nested parts of a system are shaped into the same patterns as the whole. Fractals do not define quantity but quality. This self-similarity applied to organizations can make them agile and responsive. For example, in an organization in which self-similarity of values and processes has emerged at all levels and in all geographic areas, effective teams can be assembled very quickly to take advantage of sudden opportunities or handle unexpected threats.
- In evolutionary theory, those species survive that are most capable of adapting to the environment as it changes over time. In rapidly changing global markets, the actions of one player trigger actions and reactions of other players whose actions feed back on the actions of the former. This

coevolution is the reason why companies today must run as fast as they can just to maintain their current position.

- *Group self-organization* enables a unity to emerge from individual diversity. Like individuals, work teams and organizations too can develop behavioral patterns.

Organizational Ecosystems

Like complex organisms, smart organizations have a "nervous system" which enables them to thrive on chaos and to guide them through turbulent times. Organizational nervous systems provide the functions of sensing and learning, communications—internal and external— coordination, and memory. In fast-moving, unpredictable digital environments, "nervous system" functions are essential to provide the organization with anticipatory, filtering, empathic, learning, and adaptive capabilities in real time (Por, 2000).

Second, economic activity is fractal, in that it shows the same structure and obeys the same rules for creating value at the level of the economy, the organization, and the individual. Therefore, smart organizations will need to be adaptive to their economic environment—that is, open with permeable boundaries, operating at the edge of chaos (Warnecke, 1992; Davis & Meyer, 1998).

Third, the fittest will survive. Smart organizations become fit through variety and diversity of thought, old and new ideas, that breed innovation. Cross-functional, multidisciplinary teams capable of creativity are an essential element to this. Combined with openness, through ideas from the market and interorganizational exchanges, organizational fitness grows.

Fourth, by being big and small at the same time. The essence of ecosystems is the balance between big and small organisms dependent on one another. Likewise, smart organizations must be big to afford large-scale investments, but they also must be small, nimble, unified around a purpose, and capable of paying attention to the details of important relationships (ecosystems of smart organizations, see Figure 3).

The smart organizations depicted in Figure 3 are composed of teams (dots) that are linked via ICT-enabled business processes between individuals and teams inside or outside the organization (connecting lines).

Figure 3. Ecosystem of smart organizations

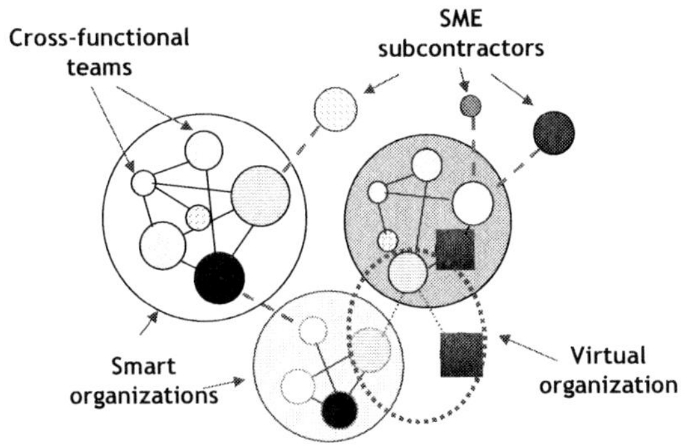

Managing the Smart Organization

Organizations in the digital age, unlike industrial age ones, will not seek to control their environments. Rather, they will adapt to them, since they recognize that any attempt to control would at best fail, and at worst stifle the creativity and imagination necessary to support innovation. In a globally networked economy, participants are free to focus and re-focus their commitment as they see fit. With this in mind, management style is evolving from one, which used to place emphasis on planning, organizing, and controlling, to one, which emphasizes providing vision, motivation, and inspiration (Kostner, 1996).

Also, in the *internet*worked economy, the roles of "superior" and "subordinate" are becoming blurred and management becomes fuzzy—that is, more laid-back, less controlling, and trust-based (Filos & Banahan, 2001a).

A bureaucracy is an efficient organizational scheme for tackling recurring tasks in a sequential way. Its static structure guarantees stability and reliability. However, team-based (networked) organizations are better able to handle tasks that are nonroutine and which demand a high degree of flexibility and adaptability. They are also able to link expertise that is distributed throughout the organization. The flexible structure of teams thus guarantees a dynamic and competent response to ad hoc tasks (Figure 4).

The organization of work in the *internet*worked economy is shifting from stable, physically colocated functions to dynamic, competence-based virtual teams that create value by synthesizing information across geographical and

Figure 4. Organizational culture of the industrial vs. the digital age

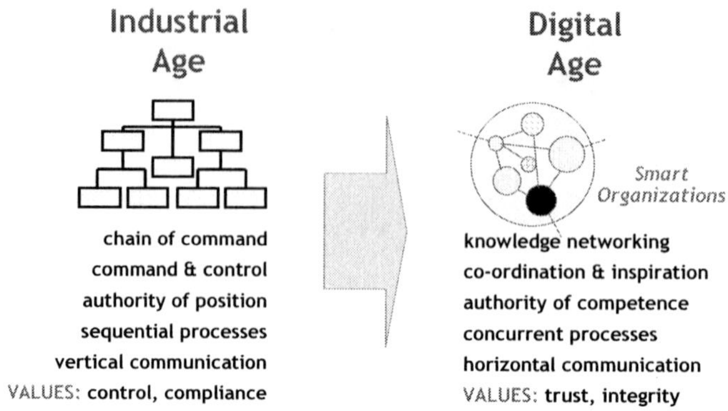

organizational boundaries. As a consequence to this, organizational culture and management change as well.

The Self-Organizing Distributed Team

The face of work is changing, too. As the business world becomes more complex due to demands for flexibility and shorter response times, the nature of work has to keep pace with organizational change. Work in smart organizations is therefore marked by concurrent work practices, flexible and versatile teamwork, and creativity and intelligent use of ICT (Figure 5).

A *virtual organization* is a collection of geographically distributed, functionally, and/or culturally diverse entities that are linked through ICT and rely on lateral, dynamic relationships for coordination (Camarinha-Matos & Afsarmanesh, 1999; Filos & Ouzounis, 2003). Despite its diffuse nature, a common identity holds the organization together in the minds of its constituents. The virtual organization is managed via teams consisting of geographically dispersed employees, forming a "company without walls," a collaborative network of people working together, regardless of location or who "owns" them (DeSanctis & Monge, 1998). A major distinction between virtual and other organizational models consists in that the former are networked (via ICT), transcend organizational boundaries (Grabowski & Roberts 1998), and should therefore be viewed as metaphors of organization design that is held together, literally, by communication.

Figure 5. The evolution of collaborative work and the impact of organizational forms

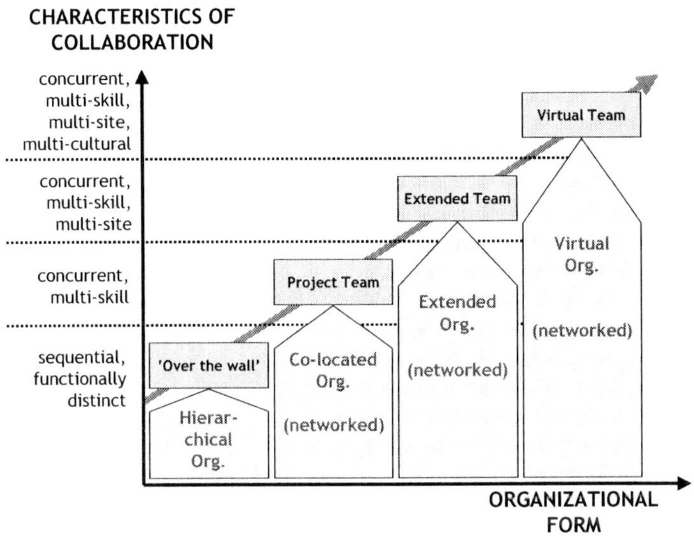

A *virtual team* is defined as a temporary, culturally, and/or functionally diverse, geographically dispersed, ICT-mediated communicating work group (Jarvenpaa & Leidner, 1998). As virtual teamwork is fast becoming a dominant way of working with many organizations, successful management of virtual teams constitutes a key component to managing virtual organizations. As virtual teams are made up of individuals with human needs for belonging, communicating, and togetherness, a radically new approach to and interdisciplinary understanding of virtual team management is required in order to harness the benefits and to develop the potential of this new socioeconomic paradigm.

Virtual Team Communication

When individuals are working together toward a common goal, the success of their undertaking depends, to a large extent, on the information exchanged, which is heavily dependent on the quality of communication between those involved. As communication between human beings involves far more than merely an exchange of information at a rational level, factors such as the emotional atmosphere, the social and cultural context, as well as nonverbal aspects may not be neglected. Contrary to earlier reservations, computer-mediated communication needs do not necessarily have a reductionistic impact on team work, but may rather contribute to "revolutionizing" its potential (Lipnack & Stamps, 1997; Devine & Filos, 2001).

In the traditional team environment, in which individuals are colocated, communication happens via conventional means, such as oral or written forms of interpersonal discourse. While written communication is almost exclusively perceived as formal and legally binding, oral communication is differentiated according to the informational settings (formal meetings or informal social events) in which it is embedded.

On the other hand, communication between individuals of remote teams has to rely almost exclusively on ICT. The distinction between the oral and the written, and with it the distinction between formal and informal discourse, may become blurred. Ong (1982) thus speaks of the "secondary orality" of the digital age. As a result, other distinctive features are likely to become important, such as ease of use, interactivity (which allows the user to feel involved), and even the noninteractivity of asynchronous communication tools (e.g., e-mail).

The Impact of Organizational Culture

The very technologies that offer individuals the freedom to work anytime and anywhere may also fray the ties that bind organization members to each other and to their employer. In particular, the cues that pull team members together in traditional organizational settings include dress codes, shared language, shared organizational culture (e.g., routines and processes), office buildings, and colocation. Consequently, since all these factors are less readily available and less indicative of meaning in the virtual context, the links between virtual team members may be less tangible, and thus more social and psychological in nature. Wiesenfeld, Raghuram, and Garud (1998), in their study on the effects of different communication media on the organizational identification of virtual workers, found that electronic media are particularly important to maintaining organizational identification due to the strong correlation of the frequency of use with it, whereas face-to-face contact may be more critical for creating it. Research on new organizational forms needs to consider the "system of work" and the "system of meaning." the institutional facets of the organization, specifically the values attached to the work engaged in (Scott, 1991). Organizational identification is a part of the larger construct that has to do with the creation and preservation of the "system of meaning" in new work forms.

Trust in the Virtual Context

Handy remarks (1995) that virtual teams are run on trust rather than on control. Indeed, the effective coordination and management of the virtual team seem to pose a real challenge. Although team cohesion may suffer from a lack of immediacy in team members' interactions due to geographical dispersion, divergence of expertise levels, or a socio-organizational heterogeneity, research results suggest that in cross-cultural virtual teams, trust takes on a form of "swift trust" that is based on clear role divisions among members who have well-defined specializations (Jarvenpaa & Leidner, 1998). Communication mediated by ICT provides the virtual platform for an informal and open sharing of thoughts, expectations, assumptions, and values. It offers an opportunity to form alliances of collective responsibility that may be different from the formal hierarchies of management relationships within the parent organizations. The virtual context may thus prove advantageous in providing clarification, sense making, and motivation for the individuals involved. This way, the value of team members' contributions is recognized and used better for the good of the community. In the end, high levels of virtually enabled trust, established between team members, may pioneer a strengthening of links between the member organizations partaking in a virtual collaboration (Grabowski & Roberts, 1998).

Leadership Conventions

Virtual teams enjoy the freedom to define for themselves the management and task assignment schemes that best suit their specific situation. Indeed, each team can build its own project culture, which can be tailored to its needs and goals, and it is certainly less "bureaucratic" than the culture in team members' organizations. Since the virtual context requires lateral communication and active involvement from each individual, it undeniably demonstrates a flat organizational structure, participatory management practices, and novel schemes of shared responsibility (e.g., management tasks performed in rotation).

In traditional teams, the focus on the team leader's role is prone to downgrade the position of the other team members. In that context, the most senior, most experienced, member is appointed as team leader. This hierarchical management scheme, as well as the assumption that teams require a single leader, is called into question in the virtual context, as teams here benefit from having different types of leaders performing complementary tasks, depending on project stage. In the virtual team, each member is empowered and responsibility is shared. Also, since there is no one person or institution to which all team members are accountable, penalties for noncompliance to the rules are imposed by the team members themselves (Jarvenpaa & Shaw, 1998).

Coping with Overabundant Information

In the digital age, the great problem may turn out not to be lack of information access but rather an overabundance of information. As Herbert Simon said, "the wealth of information creates a poverty of attention" (Shapiro & Varian, 1999). Even as passive partakers in the Information Society, people unconsciously become active contributors to this surplus of information. This is because ICT can make people vulnerable to accessing more information than they can "digest," and this can amount to a threatening drawback for organizational efficiency. Smart organizations will therefore need to manage relationships on the basis of techniques that help win the attention of people.

Weiser and Brown (1998) use human optical vision as an analogy to explain information overload and discuss possibilities to avoid it. ICT through their ubiquitous and voluminous provision of information, must engage a richer periphery. In trying to catch up with an increasing "volume of bits," users may be helplessly overwhelmed. The tools developed and used need to engage the periphery as well as the center. A balanced view must be sought continuously.

Nurturing the Knowledge Process

Managing knowledge is a core competence of the "smart" organization. In the digital economy knowledge becomes the primary raw material and result of economic activity.

The initial challenge in moving toward organizational smartness, and in order to leverage the power of knowledge, one must know where to find it and once found, know what to do with it. Knowledge can be either explicit or tacit (Polanyi, 1966). In the case of the former, knowledge is formal and systematic and thus easy to capture, store, and communicate. Tacit knowledge on the other hand is personal, a combination of experience and intuition, and as such the organization's ability to capture and communicate it is heavily dependent on the individual owner's commitment to the organization and to its need to generate value from it. In this sense, a great deal of trust and loyalty between the individual and the organization is necessary to leverage organizational knowledge, including its tacit dimension.

It is therefore essential to make a proper distinction between the terms "data," "information," and "knowledge" (Figure 6). The interchangeable use of "information" and "knowledge" tends to obscure the fact that while it can be easy and quick to transfer information from one place to another, it may often involve a very difficult and slow process to transfer knowledge. Knowledge is a human capability that can be acquired and expanded through learning. In trying to define knowledge it can be helpful to realize that the human mind is considered capable of two kinds of knowledge, the rational and the intuitive.

In Western thinking, intuitive knowledge has been devalued in favor of rational scientific knowledge. In Eastern thinking however, the tradition has been to recognize the importance of the intuitive. Chinese philosophy has emphasized the complementary nature of the intuitive and the rational and has represented them by the archetypal pair yin and yang.

Recognition of the difficulties inherent in transferring knowledge from one person to another has tended to highlight the importance of tacit knowledge. This heuristic, subjective, and internalized knowledge is not easy to communicate and is learned through practical examples, experience, and practice. Where explicit, articulate knowledge may be stored in the form of a patent or as documented know-how, tacit, nonarticulate knowledge is communicated in

Figure 6. The knowledge pyramid: A tentative approach to explaining "slippery" terms

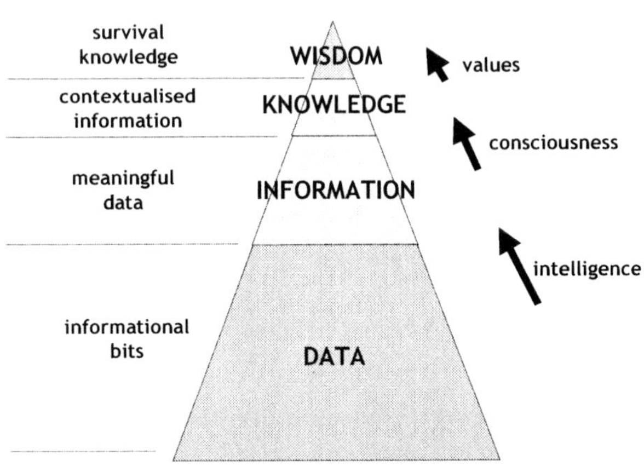

social networks, or know-who. Debates over the meaning of knowledge are ongoing, and do not seem likely to end for some time to come. Similarly, there is no agreed definition of knowledge management. The term is used loosely to refer to a broad collection of organizational practices and approaches related to generating, capturing disseminating know-how, and other content relevant to an organization's business. Knowledge is thus not an explicit, tangible "thing", but information combined with experience, context, interpretation, and reflection. Also, knowledge involves the full person, integrating the elements of both thinking and feeling. Knowledge management is thus increasingly seen as signaling the development of a more organic and holistic way of understanding and exploiting the role of knowledge in the process of managing and doing work, and an authentic guide for individuals, teams, and organizations in coping with the increasing complexity of modern business environments.

Stewart (1998) uses the term "intellectual capital" to denote intellectual material—knowledge, information, intellectual property, experience—that can be put to use to create wealth. Intellectual capital is to be seen as an asset for every organization residing in its people (human capital), its structures (structural capital), and its customers (customer capital).

Achieving and Maintaining a High Level of Creativity

Smart organizations embody cross-functional, multidisciplinary teams. Their creativity is based on knowledge networking within and across the organization's boundaries. This openness to ideas drives the creativity of the whole organization.

As Toffler (1981) illustrates, the distinction between producer and consumer diminishes as consumers begin to play an important role, such as in the development or the further improvement of a product. Mass customization enables smart organizations to see customers, suppliers, regulators, and even competitors as stakeholders with meaningful contributions.

Redundancy Frameworks

Building redundancy is a way to support creative teams. Redundancy comes from intensive communication on a common cognitive ground and the facilitation of tacit knowledge transfer. While team members share "overlapping" information, they can sense what others are struggling to articulate.

One way to achieve redundancy is to organize teams in competition with each other. In one sense, such internal competition is wasteful. However, when responsibilities are shared, information proliferates, and the organization's ability to develop and implement efficient concepts is accelerated.

Another way to enable redundancy is through rotation, especially between different functions such as R&D and marketing. Rotation helps employees understand the business from a multiplicity of perspectives. Changing roles and responsibilities helps create and maintain team spirit and commitment to the team objectives, but most importantly, it may drive innovation within the team, as a result of augmented lateral thinking and knowledge sharing.

Active Knowledge Sharing

Metes, Gundry, and Bradish (1997) propose a computer-mediated approach to facilitating knowledge management and creativity of distributed teams. The tool they propose is computer conferencing, also known as "chat" tool. They argue that in contrast to teams using the telephone, fax, e-mail, or audio and video conferencing, teams that use computer conferencing create a permanent shared record of their communication. This is specifically important because information is transmitted in its proper contextual setting, including situations, relationships, assumptions, expectations, and history. Adding context to information transforms it into knowledge (see Figure 6).

Investments in intellectual assets, unlike investments in physical assets, increase in value with use. Properly stimulated, knowledge and intellect grow exponentially when shared. If two people exchange knowledge with each other, both gain information and experience. And if both then share their new knowledge with others—each of whom feeds back questions, suggestions, and modifications—the benefits can grow exponentially. Once an organization gains a knowledge-based competitive edge, it becomes ever easier for it to maintain its lead and ever harder for competitors to catch up.

Professional intellect (Nonaka & Takeuchi, 1995) of an organization operates on four levels:

- Cognitive knowledge (know-what), the basic mastery of a discipline, achieved through extensive training and certification.
- Advanced skills (know-how), the ability to apply the rules of a discipline to complex, real-world problems.

- Systems understanding (know-why), a deep knowledge of the interlinked cause-and-effect relationships underlying a discipline.
- Self-motivated creativity (care-why), which consists of will, motivation, and adaptability for success. Here lies the reason why highly motivated creative teams often outperform teams with greater physical or financial resources. This level depends on the organizational culture.

The value of professional intellect increases when moving up the intellectual scale from cognitive knowledge to self-motivated creativity. Unfortunately, most organizations focus their training efforts on developing basic skills and only very few invest in developing systems and creative skills.

Research on the Smart Organization

Research and development (R&D) has contributed substantially to the emergence of smart organizations. In Europe for example, successive research framework programs in the last 12 years have supported the development of technologies that facilitated electronic commerce and digital business. In the early 1990s, research focused on concurrent engineering (Fan & Filos, 1999), on computer-supported collaborative work and product and process data modeling. The work program of the European Strategic Program in Information Technologies (ESPRIT,1997), in the domains high-performance computing and networking, technologies for business processes, and integration in manufacturing supported R&D relevant to the virtual enterprise. Between 1994 and 1998, more than 50 industry-led projects were set up with around 100 million-Euro funding (shared cost funding with 50% industrial contribution). In addition to regular consultations with industry, a number of projects were established which brought together major industrial users of information technology (IT) and the vendor community. The common aim of these projects was to set long-term research targets for the IT industry in order to meet well-formulated industrial needs. The Advanced Information Technology initiative, for example, dealt with the automotive and aerospace industries (AIT, 2001). It comprised 22 R&D projects that also had a major impact on standardization developments. All these projects were operating concurrently within a harmonization framework (Garas & Naccari, 2001). Forty percent of organizations participating in ESPRIT were industrial user enterprises. In total, 65% of participants in ESPRIT were industrial companies. Until 1999, R&D support for the "virtual enterprise" in Europe was mainly through ESPRIT and its international cooperation activities under the Intelligent Manufacturing Systems (IMS, 2005) framework.

In 1999, the Information Society Technologies program (IST, 1999) emerged as an integrated program from previous programs ESPRIT, Advanced Communications Technologies and Systems (ACTS, 1998), and Telematics (TAP, 1998). In the work program of IST, the perspective had changed from "virtual enterprise" to any type of "virtual organization". Under the new program's Key Action II (New Methods of Work and Electronic Commerce), several calls for collaborative research proposals were launched under topics such as "dynamic networked organizations", "smart organizations", and "dynamic value constellations". In parallel, research in learning and cognition had led to the introduction of a new research field, "organizational knowledge management".

All these R&D efforts have contributed to a strong research foundation for the development of smart organizations in Europe (Filos & Ouzounis, 2003; Wagner et al., 2004).

Research on the Virtual Organization

In parallel to these European research activities, research relevant to the virtual organization in the United States was undertaken mainly under defense contracts funded by the Defense Advanced Research Projects Agency (DARPA) and through grants of the National Institute for Standards and Testing and the National Science Foundation (Goranson, 1999).

Between 1999 and 2002, under the European IST program, more than 200 R&D projects were launched on organizations research and on research in e-business and e-work, with a total funding of about 450 million Euro. These fall into three subareas: ICT; work, business and organizational aspects; and socioeconomic issues (Zobel & Filos, 2002; Filos, 2005; Camarinha-Matos et al., 2005).

ICT Aspects of Virtual Organizations

The part of the project portfolio dealing with activities related to the design and development of generic infrastructures to support collaborative business in a networked environment involved issues such as safe communications, interoperability and tools integration, information and knowledge sharing, repositories, coordination mechanisms, and collaborative environments. These projects worked towards the emergence of a general "plug-and-do-business" architecture for interoperability (Bacquet & Naccari, 2002; Doumeingts & Chen, 2003). Project GLOBEMEN aimed at creating an IT infrastructure to support globally distributed and dynamically networked operations in one-of-a-kind industries (Karvonen et al., 2003), COMMA and BUSINESS ARCHITECT made extensive use of modeling and knowledge sharing to support virtual enterprise process integration.

As far as the characteristics and requirements regarding interoperability and information exchange are concerned, innovative approaches were required. Interoperability was to become a "design principle" while aiming to preserve the diversity, autonomy, and heterogeneity of components and environments. For example, project ECOLNET sought to validate different business strategies for independent small- and medium-sized enterprises (SME) focusing on their national market, E-COLLEG investigated an infrastructure to establish a backbone for collaborative engineering (Witczynski & Pawlak, 2002), CO-OPERATE focused on coordination of manufacturing, planning, and control activities in supply chain management, and WHALES developed a planning and management infrastructure for distributed organizations working as networks on large-scale engineering projects.

The projects portfolio was strong in demonstrating the feasibility of operating the virtual organization. The technologies used involved the Java framework, CORBA, XML, Web services, multi-agents, and modeling tools based on UML. The general aim was to use standards whenever possible. This aspect is particularly clear with respect to de facto standards being proposed by industry groups such as the Object Management Group, the Workflow Management Coalition, the World Wide Web Consortium (W3C), and the UN Center for Trade Facilitation and Electronic Business (ebXML).

The significance of virtual organization modeling and interoperability of applications arose from the need to model the virtual organization as a means to properly understand and manage it. A problem with existing business process modelers lies in how to translate one model based on one proprietary modeling technique into an equivalent model represented by another. One strategy pursued in Europe was in agreeing on a basic language that makes such transformations possible. Consensus was reached and the Unified Enterprise Modeling Language was defined (UEML, 2004).

Some projects dealt with ontologies, conceptual information models that describe things that exist in a domain, whose purpose was

- To support human understanding and organizational communication.
- To be machine-processable and thus facilitate content-based access, and communication, and integration across different information systems.

A decade of international research has led to the creation of ontology languages, editors, reasoning techniques, and development guidelines. Various languages for ontology specification and implementation are now available. These languages have built-in reasoning techniques, and they also allow developing special purpose reasoning services.

An area of impact is the Semantic Web, in which computers "find the meaning" of data in automated Web services such as functional agents. The DARPA Agent Markup Language (DAML) and the Ontology Inference Layer (OIL) that was developed by the World Wide Web Consortium and the European OIL community (W3C, 2001), provide a rich set of constructs with which to create ontologies and to mark up information so that it becomes machine-readable. A significant number of European projects addressed knowledge technologies in the context of the virtual organization and business collaboration (Filos, 2002).

Work, Business, and Organizational Issues

This subarea involved reference models and architectures, such as the specification of logical reference architectures for new and emerging cooperative organizations by identifying the main functional blocks, interactions, actors and their roles, resources, and value systems, as well as the definition and the characterization of collaborative business models, the forms of cooperation in networked environments and means to assess the effectiveness of virtual organizations. Work involved virtual organization reference models, collaborative business models (and related case studies), cooperation methodologies and performance measurement. The projects addressed centralized support services as well as services that are distributed across the virtual organization (Hartel, Sonderegger, Kamio, & Zhou, 2002; Kazi, Hannus, & Ollus, 2002; Katzy & Sung, 2003).

Some projects addressed business functions of the various parts of the life cycle of a virtual organization. Research activities included partner registration and search, marketplace management, e-procurement and negotiation, distributed business process planning and management, and so forth, with a particular focus on domain-independent services covering the various phases of the life cycle of a virtual organization. They also comprised supervision and monitoring, as well as specialized services, such as contract modeling and negotiation, a support infrastructure to help virtual enterprises to address the legal issues involved, as well as a Web-based infrastructure for alternative online dispute resolution for SME (Gouimenou, 2001).

Through its IST program, the European Commission also supported a range of projects that aimed to accelerate e-business technology take-up in SME. The concept behind these projects was to transfer leading-edge technologies to industry and other end-users. Under Key Action II, between 1998 and 2002, more than 70 take-up projects were launched, which demonstrated the relevance of e-business, e-commerce, and e-work technologies for SME. Hundreds of SME throughout Europe participated together with so-called "catalysts"—local or regional organizations that worked with them to help them adapt their business processes toward better ICT use. The SME were able to "rethink" and adapt emerging technologies to their business needs by sharing development effort and jointly achieved results among one another. These take-up projects thus became a means to leverage the results of IST research and to contribute to the implementation of the European Commission's eEurope (2005) initiative at local level, by supporting SME directly or indirectly.

The 70 million Euro invested in this take-up project's pilot activity represent only a small fraction of the total European investment in e-business. They were essential, however, in demonstrating that investment in R&D and technology transfer can be a useful instrument to help increase SME competitiveness in today's global market places. Twenty-two showcases are presented in a book (eBiz, 2003). They complement European Member States' efforts, such as those under the GoDigital initiative (2002).

The Socioeconomic Perspective

Between 1999 and 2002, socioeconomic research within IST was a significant nontechnological research activity that aimed at complementing technology activities. It was implemented through a series of calls for proposals. The primary scope of this research was in methods and tools and in understanding the impact of ICT on the economy and on society at large. The main beneficiaries were the program's research community, industry, and policy makers (Hayfa & Filos, 2003). More than 40 projects addressed socio-organizational or socioeconomic issues: industrial and organizational aspects of the digital economy (e-business, e-work), as well as societal aspects; e-business models and intangible assets; impact assessment, mainly at microlevel; corporate social responsibility; statistical indicators. Also, a number of key legal and regulatory issues emerged as a result of this research activity. Some of them were explicitly addressed; for example, legal aspects of virtual enterprises, contract law (intra-/inter-organizational or that of individuals), alternative dispute resolution, digital rights management, intellectual property rights, consumer protection, and related legal aspects (Merz et al., 2001; Hassan, Carter, Seddon, & Mangini, 2001; Van Schoubroeck, Cousy, Droshout, & Windey, 2001; Carter, 2002). All these activities contributed to the definition of a virtual organizations framework (Camarinha-Matos et al., 2004).

Research in Knowledge Management

The European Commission has supported research in knowledge management since the late 1980s, long before knowledge management itself was a recognized term. Early contributions were made in areas such as information management, quality management, and the social sciences. The first formal initiative was launched in 1998 under the research theme "Learning and Training in Industry" (LTI), as part of the ESPRIT program. Under the LTI initiative 16 research projects were launched involving more than 100 research and user organizations. Although the situation has evolved considerably since then, many current projects have their roots in this initial incursion into the realities of organizational learning.

As knowledge management concepts and practices caught the attention of organizations across Europe, European-funded research moved squarely toward supporting the development of solutions that enable individuals to share knowledge within and among organizations as part of the innovation process. The main focus of research has been on supporting multidisciplinary solutions and practices for individuals and corporations to manage knowledge within

Table 1. Knowledge management research in the IST (2002) program

Projects funded under the IST Program reflect a broad spectrum of KM approaches and theories. They can be classified broadly as follows:
First Generation KM
Information portals—tools and methodologies integrating to a greater or lesser extent information necessary for back and front office processes in organizations. These projects mainly originated from the first call for proposals in IST (1999).
Second Generation KM
Knowledge processes to business processes—tools and methodologies linking knowledge and business processes
Assessment or measurement-type projects—which attempt to measure and benchmark knowledge management implementation within and between organizations and to manage and measure impact of knowledge lifecycles within the enterprise
Collaboration and innovation spaces—tools, methodologies and good practices to accelerate creative exchanges among people working within and across organizations. The end objective of such projects is to support the transition of organizations into knowledge-based communities.
Third Generation KM
Knowledge and innovation ecologies—tools, methodologies and good practices which identify contextual barriers and enablers of absorptive and innovative capacities of organizations and attempt to replicate co-creation abilities across the enterprise or network
Human-centered knowledge management—focus on people as unique holders of knowledge, and exchanges between people as primary generators of new knowledge for innovation.
Networks and working groups—which attempt to build critical mass within and outside the IST program.

networked organizations and communities of practice. Specifically, this included aspects such as:

- Integrated ICT platforms, including mobile, to manage the full lifecycle of knowledge (i.e., its capture, organization, maintenance, mining, sharing and trading) in support of both intra- and inter-organizational activities
- Personalized, context-, task- and role-sensitive functionality for the dynamic provision and sharing of timely and relevant knowledge.
- Solutions to organize and exploit heterogeneous, unstructured information sources, using ontologies, self-organization paradigms as well as semantic cross-lingual search, in support of e-work and e-commerce applications.
- Tools and environments for knowledge sharing, collaboration, and socialization within and among organizations which build on methodologies from areas such as organizational behavior, cognitive psychology, human factors, man-machine dialogue, as well as social and management sciences.

The research activities focused not just on technology development but also on its application. In addition to R&D projects, the European Commission also funded a variety of take-up and support activities designed to help make knowledge management better known and accepted notably in small and medium enterprises.

Under the IST program, the "Knowledge Management Made in Europe" (KMME) initiative was launched after the start of the Fifth Framework Program (1999-2002), with an aim to "create a strong brand for European KM research and practice" and to "bring into the portfolio quality proposals". The overall goal of the initiative at the outset was to increase European competitiveness, to improve the working life of European individuals, and to build on European strengths of languages, cultural diversity, and industrial leadership.

One of the major epistemological directions the initiative declared was to pursue the challenge of complexity as a key factor in the knowledge economy, using a holistic approach. The initiative involved 58 research, take-up, and cluster projects with a total public investment of approximately 65 million Euro. Projects funded fell into the three categories outlined in Table 1.

The first category, or first set of projects to be funded, were denoted "first generation KM" (under LTI in 1998) and concentrated on themes and concepts such as information portals, tools and methodologies integrating to a large or lesser extent information necessary for back and front office processes in organizations.

The second wave, from 1999 to 2000, aimed at a more holistic treatment of primarily tacit knowledge in organizations and funded projects with concepts and themes such as linking knowledge processes to business processes, assessing KM implementation and collaboration and innovation spaces.

The third generation KM (2001-2002) represented a movement away from the classical knowledge management engineering approach, and aimed at funding projects with concepts and themes such as knowledge and innovation ecologies and human-centered KM.

One of the most conspicuous and most mentioned projects with the largest international profile is the European Knowledge Management Forum (EKMF), a cluster project which attempted to "build a sustainable network of Knowledge Management theoreticians and practitioners who are interested in Europe's journey into the knowledge economy, and what Knowledge Management methods and tools can contribute to this journey." (KnowledgeBoard, 2005).

An assessment of the KMME initiative (Sage, Stanbridge, & Shelton, 2004) to date shows that projects funded in the first wave are indicating a focus on classical, engineering approaches to knowledge management. This concentration is typical of early projects in knowledge management programs. The same phenomenon was observed in the U.S. in knowledge management research. Many of the projects in the first phase were industry- or sector-specific and helped to solve problems specific to the sector or industry, without addressing issues that were of benefit to different sectors or with impact on the industry value chain.

The second wave marks a shift from the engineering approach to a more centralist, best-practice approach. In the third wave, a significant number of projects were funded that are advanced on the mathematical complexity scale and address concepts such as intelligent agents and the Semantic Web.

However, only a few projects address the area of social complexity, which has high potential for KM that is related to the European context of linguistic and cultural diversity. The subject of complexity is not widely recognized within the KnowledgeBoard community.

The phenomenon of divergence between focus areas in knowledge management research in Europe, and a false dichotomy between human-centered approaches and engineering/mathematical approaches has been observed in the U.S. as well. The opportunity for Europe is to fund and initiate more research that is related to the human-centered approach, but also looks at social complexity.

Conclusion

This chapter aimed to draw a picture of the changing organizational paradigm in the digital age. Successive European R&D programs played a significant part in developing the technologies and concepts that are key to those developments. The research efforts aimed at understanding and improving knowledge management, the virtual organization and digital business processes. While many of the features of digital age organizations are not yet fully understood, there is hope that organizations in the future will become "smart" in various respects. The unprecedented opportunities offered by Information Society for individuals to relate with one another, to work, and to do business in digital environments will change the ways organizations relate to each other and to the individuals that are key to their core competences.

Acknowledgment

The views expressed in this paper are those of the author and do not necessarily reflect the official view of the European Commission on the subject.

References

ACTS (1998). *Advanced Communications Technologies and Services program.* European Commission, Directorate-General XIII—Telecommunications, Brussels. Retrieved from http://www.cordis.lu/acts/

AIT (2001). Advanced Information Technology (Garas & Naccari, 2001). Similar initiatives for the process industries (PRIMA) and the construction (or large-scale engineering) industries (ELSEWISE, see Mitrovic et al., 1999; McAffer & Garas, 1999). Of similar impact for the furniture industry was FUNSTEP (Jardim-Gonçalves et al., 1999).

Bacquet, J., & Naccari, F. N. (2002). "Plug and do business" and the European R&D programs. In L. M. Camarinha-Matos (Ed.), *Collaborative business ecosystems and virtual enterprises, IFIP TC5/WG5.5 Third Working Conference on Infrastructures for Virtual Enterprises (PRO-VE'02)* (pp. 283-292). Boston: Kluwer Academic Publishers.

Camarinha-Matos, L. M., & Afsarmanesh, H. (1999). The virtual enterprise concept. In L. M. Camarinha-Matos & H. Afsarmanesh (Eds.), *Infrastructures for virtual enterprises: Networking industrial enterprises* (pp. 3-30). Boston: Kluwer Academic Publishers.

Camarinha-Matos, L. M., & Afsharmanesh, H. (Eds.). (2004). *Collaborative networked organizations: A research agenda for emerging business models*. Boston; Dordrecht; London: Kluwer Academic Publishers.

Camarinha-Matos, L. M., Afsharmanesh, H., & Ollus, M. (Eds.). (2005). *Virtual organizations. systems and practices*. New York: Springer Science + Business Media, Inc.

Camarinha-Matos, L. M., Afsharmanesh, H., & Rabelo, R. J. (Eds.). (2001). *E-business and virtual enterprises: Managing business-to-business cooperation*. Dordrecht: Kluwer Academic Publishers.

Carter, C. (2002). The impact of contractual support for ICTs on working practices. In B. Stanford-Smith, E. Chiozza, & M. Edin (Eds.), *Challenges and achievements in e-business and e-work* (pp. 1291-1298). Amsterdam: IOS Press.

Cunningham, P., Cunningham, M., & Fatelnig, P. (Eds.). (2003). *Building the knowledge economy: Issues, applications, case studies*. Amsterdam: IOS Press.

Davis, S., & Meyer, C. (1998). *Blurr—The speed of change in the connected economy*. Oxford: Capstone.

Deloitte & Touche. (2002). *Collaborative knowledge networks*. New York: Deloitte Consulting and Deloitte & Touche.

DeSanctis, G., & Monge, P. (1998). Communication processes for virtual organizations. *Journal of Computer-Mediated Communication, 3*(4). Retrieved from http://jcmc.indiana.edu/vol3/issue4/desanctis.html

Devine, M., & Filos, E. (2001). Virtual teams and the organizational grapevine. In L. M. Camarinha-Matos, H. Afsharmanesh, & R. J. Rabelo (Eds.), *IFIP TC5/WG5.3 Second Working Conference on Infrastructures for Virtual Organizations: Managing Cooperation in Virtual Organizations and Electronic Business towards Smart Organizations* (pp. 413-424).

Doumeingts, G., & Chen, D. (2003). Interoperability development for enterprise applications and software. In P. Cunningham, M. Cunningham, & P. Fatelnig (Eds.), *Building the knowledge economy: Issues, applications, case studies* (pp. 207-214). Amsterdam: IOS Press.

Doz, Y. L., & Hamel, G. (1998). *Alliance advantage: The art of creating value through partnering*. Boston: Harvard Business School Press.

eBiz (2003). *European e-business Showcases, European Commission, Office for Official Publications, Luxembourg*. Retrieved from http://europa.eu.int/information_society/ecowor/smes/

eEurope (2005). eEurope 2005—*An information society for all*. European Commission, Directorate-General Information Society and Media,

Brussels. Retrieved from http://europa.eu.int/information_society/eeurope/2005/

ESPRIT (1997). European Strategic Program for Research in Information Technologies, Building the Information Society. Work Program 1997, European Commission, Directorate-General III—Industry, Brussels. Retrieved from http://www.cordis.lu/esprit/

Evans, S., & Roth, N. (2004). Collaborative knowledge networks. In L. M. Camarinha-Matos & H. Afsharmanesh (Eds.), *Collaborative networked organizations: A research agenda for emerging business models* (pp. 289-312). Boston, Dordrecht, London: Kluwer Academic Publishers.

Fan, I.-S., & Filos, E. (1999). Concurrent engineering: Esprit-supported R&D projects in a world-wide context. In N. Wognum, K. -D. Thoben, & K. S. Pawar (Eds.), *ICE'99, International Conference on Concurrent Enterprising* (pp. 177-189). Nottingham: University of Nottingham.

Filos, E. (2002). European collaborative R&D projects related to the "smart organization." A first evaluation of activities and implications for construction. In Z. Turk & R. Scherer (Eds.), *eWork and eBusiness in architecture, engineering and construction* (pp. 27-32). Lisse: Sweets and Zeitlinger..

Filos, E. (2005). Virtuality and the future of organizations. In G. Putnik & M. M. Cunha (Eds.), *Virtual enterprise integration: Technological and organizational perspectives*. Hershey, PA: Idea Group.

Filos, E., & Banahan, E. (2001a). Will the organization disappear? The challenges of the new economy and future perspectives. In Camarinha-Matos et al. (Eds.), *IFIP TC5/WG5.3 Second Working Conference on Infrastructures for Virtual Organizations: Managing Cooperation in Virtual Organizations and Electronic Business towards Smart Organizations*, December 4-6, 2000, Florianopolis, Santa Catarina, Brazil (pp. 3-20).

Filos, E., & Banahan, E. (2001b). Towards the smart organization: An emerging organizational paradigm and the contribution of the European RTD programs. *Journal of Intelligent Manufacturing, 12*(2), 101-119.

Filos, E., & Ouzounis, V. K. (2003). Virtual organizations: Technologies, trends, standards and the contribution of the European R&D programs. *International Journal of Computer Applications in Technology, 18*(1-4), 6-26.

Foy, N. (1980). *The ying and yang of organizations.* Grant McIntyre.

Garas, F. K., & Naccari, F. N. (2001). Assessment of the AIT initiative. Report to the European Commission, Directorate-General Information Society, Brussels. Retrieved from http://www.cordis.lu/ist/ka2/rptspolicyconf.htm

GoDigital (2002). The 'GoDigital' initiative was an integral part of eEurope 2002. Retrieved from http://europa.eu.int/information_society/topics/ebusiness/godigital/index_en.htm

Goldman, S. L., Nagel, R. N., & Preiss, K. (1995). *Agile competitors and virtual organizations: Strategies for enriching the customer*. New York: Van Nostrand Reinhold.

Goranson, T. H. (1999). *The agile virtual enterprise: Cases, metrics, tools*. Westport-London: Quorum.

Gouimenou, J. (2001). E-Arbitration-T: An alternative dispute resolution for SMEs. In B. Stanford-Smith & E. Chiozza (Eds.), *E-work and e-commerce: Novel solutions and practices for a global networked economy* (pp. 526-531). Amsterdam: IOS Press.

Grabowski, M., & Roberts, K. H. (1998). Risk mitigation in virtual organizations [Electronic version]. *Journal of Computer-Mediated Communication, 3*(4).

Handy, C. (1995, May-June). Trust and the virtual organization. *Harvard Business Review*, 25-50.

Handy, C. (1999). *Understanding organizations* (4th ed.). London: Penguin.

Hartel, I., Sonderegger, S., Kamio, Y., & Zhou, M. (2002). The information roadmap of virtual enterprises. In B. Stanford-Smith, E. Chiozza, & M. Edin (Eds.), *Challenges and achievements in e-business and e-work* (pp. 1380-1387). Amsterdam: IOS Press.

Hassan, T., Carter, C., Seddon, C., & Mangini, M. (2001). e-LEGAL: Dismantling the legal barriers to ICT up-take in virtual enterprises. In B. Stanford-Smith & E. Chiozza (Eds.), *E-work and E-commerce: Novel solutions and practices for a global networked economy* (pp. 602-608). Amsterdam: IOS Press.

Hayfa, T., & Filos, E. (2003). How ICTs impact policy-making: Thirteen IST project examples. European Commission, Directorate-General Information Society, Brussels. Retrieved from ftp://ftp.cordis.lu/pub/ist/docs/dir_c/rd_report_en.pdf

Imai, K., & Itami, H. (1984). Interpenetration of organization and market. *International Journal of Industrial Organization, (2),* 285-310.

IMS (2005). Intelligent Manufacturing Systems initiative. Retrieved from http://www.ims.org

IST (1999). Information Society Technologies. Work Program 1999, European Commission, Directorate-General Information Society, Brussels. Retrieved from http://www.cordis.lu/ist/library.htm

IST (2002). Retrieved from http://www.cordis.lu/ist/ist-fp5.html

Jardim-Gonçalves, R., Sousa, P. C., Pimentão, J. P., & Steiger-Garcão, A. (1999). Furniture commerce electronically assisted by way of a standard-based integrated environment. In N. Wognum, K.-D. Thoben, & K. S. Pawar (Eds.), *ICE'99, International Conference on Concurrent Enterprising* (pp. 129-136). Nottingham: University of Nottingham.

Jarvenpaa, S. L., & Leidner, D. E. (1998). Communication and trust in global virtual teams [Electronic version]. *Journal of Computer-Mediated Communication, 3*(4).

Jarvenpaa, S. L., & Shaw, T. R. (1998). Global virtual teams: Integrating models of trust. *Proceedings VO-Net Workshop* (pp. 35-51).

Karvonen, I., van den Berg, R., Bernus, P., Fukuda, Y., Hannus, M., Hartel, I., et al.(Eds.). (2003). *Global Engineering and Manufacturing in Enterprise Networks. GLOBEMEN. VTT Symposium* [Electronic version].

Katzy, B. R., & Sung, G. (2003). State-of-the-art virtual organization modeling. In P. Cunningham, M. Cunningham, & P. Fatelnig (Eds.), *Building the knowledge economy: Issues, applications, case studies* (pp. 959-966). Amsterdam: IOS Press.

Kazi, A. S., Hannus, M., & Ollus, M. (2002). Layered architecture for inter-enterprise collaboration. In B. Stanford-Smith, E. Chiozza, & M. Edin (Eds.), *Challenges and achievements in e-business and e-work* (pp. 1373-1379). Amsterdam: IOS Press.

Kelly, K. (1999). *New rules for the new economy.* London: Forth Estate.

Kelly, S., & Allison, M. A. (1999). *The complexity advantage: How the science of complexity can help your business achieve peak performance.* New York: BusinessWeek Books, McGraw-Hill.

KnowledgeBoard (2005). The Web site of the European Knowledge Management Forum (EKMF). Retrieved from http://www.knowledgeboard.com/

Kostner, J. (1996). *Virtual leadership: Secrets from the round table for the multi-site manager*. New York: Warner Books, Inc.

Levine, R., Locke, C., Searls, D., & Weinberger, D. (2000). *The Cluetrain Manifesto: The end of business as usual*. Cambridge, MA: Perseus.

Lipnack, J., & Stamps, J. (1997). *Virtual teams: Reaching across space, time, and organizations with technology*. New York: John Wiley & Sons.

McCaffer, R., & Garas, F. (Eds.). (1999). eLSEwise: European Large Scale Engineering Wide Integration Support Effort [Special Issue]. *Engineering Construction and Architectural Management*, *6*(1).

Merz, M. et al. (2001). Electronic contracting in the construction industry. In B. Stanford-Smith & E. Chiozza (Eds.), *E-work and e-commerce: Novel solutions and practices for a global networked economy* (pp. 595-601). Amsterdam: IOS Press.

Metes, G., Gundry, J., & Bradish, P. (1997). *Agile networking: Competing through the Internet and Intranets*. Upper Saddle River, NJ: Prentice Hall.

Mills, D. Q. (1991). *Rebirth of the corporation*. New York: John Wiley & Sons.

Mintzberg, H. (1980). *Mintzberg on management: Inside our strange world of organizations*. New York: The Free Press.

Mitrovic, D., Hunter, I., & Male, S. (1999). Characteristics of networked enterprise in global construction. In N. Wognum, K.-D. Thoben, & K. S. Pawar (Eds.), *ICE'99, International Conference on Concurrent Enterprising* (pp. 447-454). Nottingham: University of Nottingham. See also McAffer & Garas (1999).

Nonaka, I., & Takeuchi, H. (1995). *The knowledge-creating company: How Japanese companies create the dynamics of innovation*. New York: Oxford University Press.

O'Mahony, M., & van Ark, B. (2003). *EU productivity and competitiveness: An industry perspective. Can Europe resume the catching-up process?* Luxembourg: European Commission.

O'Reilly, C. A., & Tushman, M. L. (2004). The ambidextrous organization. *Harvard Business Review*, April, 74-81.

OECD (2003). *The policy agenda for growth: An overview of the sources of economic growth in OECD countries*. Paris: OECD.

Ong, W. J. (1982). *Orality and literacy: The technologising of the word.* London: Methuen.

Polanyi, M. (1966). *The tacit dimension*. London: Routledge & Kegan Paul.

Por, G. (2000). Knowledge, intelligence, wisdom: The new economic value chain and its enabling technologies. In Future Organizations and Knowledge Management, Program Consultation Meeting, European Commission, Directorate-General Information Society, Brussels.

Sage, J., Stanbridge, P., & Shelton, R. (2004). *Business knowledge management: A study on market prospects, business needs and technological trends*. Final Report for the European Commission, DG Information Society, IBM Business Consulting Services. Retrieved from http://europa.eu.int/information_society/doc/library/business_knowledge_management.pdf

Savage, C. M. (1996). *5th generation management: Co-creating through virtual enterprising, dynamic teaming, and knowledge networking.* Newton, MA: Butterworth-Heinemann.

Schapiro, C., & Varian, H. R. (1999). *Information rules: A strategic guide to the network economy.* Boston: Harvard Business School Press.

Scott, W. R. (1991). Unpacking institutional arguments. In W. W. Powell & P. J. DiMaggio (Eds.), *The new institutionalism in organizational analysis* (pp. 164-182). Chicago: University of Chicago Press. Cit. Wiesenfeldt, B. M. et al. (1998).

Senge, P. M. (1990). *The fifth discipline: The art and practice of the learning organization*. London: Random House.

Slouka, M. (1995). *War of the worlds: Cyberspace and the high-tech assault on reality*. New York: Basic Books.

Soete, L. (Ed.). (1996). *Building the European information society for us all: First reflections of the high level group of experts*, Interim Report, European Commission, Directorate General V, Brussels.

Stanford-Smith, B., Chiozza, E. (Eds.). (2001). *E-work and e-commerce: Novel solutions and practices for a global networked economy.* Amsterdam: IOS Press.

Stanford-Smith, B., Chiozza, E., & Edin, M. (Eds.). (2002). *Challenges and achievements in e-business and e-work*. Amsterdam: IOS Press.

Stewart, T. A. (1998). *Intellectual capital: The new wealth of organizations*. London: Nicholas Brealy Publishing.

Stoll, C. (1995). *Silicon snake oil: Second thoughts on the information superhighway*. New York: Doubleday.

TAP (1998). Telematics Applications Program. European Commission, Directorate-General XIII - Telecommunications, Brussels. For further information see: http://www.cordis.lu/telematics/

Tapscott, D., Lowy, A., & Ticoll, D. (Eds.). (1998) *Blueprint to the digital economy: Creating wealth in the era of e-business*. New York: McGraw Hill.

Ticoll, D., Lowy, A., & Kalakota, R. (1998). Joined at the bit: The emergence of the e-business community. In D. Tapscott, A. Lowy, & D. Ticoll (Eds.), *Blueprint to the digital economy: Creating wealth in the era of e-business* (pp. 19-33). New York: McGraw Hill.

Toffler, A. (1981). *The third wave*. London: Pan Books.

UEML (2004). Retrieved from http://www.cimosa.de/Modelling/UEML02.html and http://www.rtd.computas.com/websolution/Default.asp?WebID=239

Van Schoubroeck, C., Cousy, H., Droshout, D., & Windey, B. (2001). Virtual enterprise legal issue taxonomy. In B. Stanford-Smith & E. Chiozza (Eds.), *E-work and e-commerce: Novel solutions and practices for a global networked economy* (pp. 609-615). Amsterdam: IOS Press.

W3C (2001). World-Wide Web Consortium, Web Ontology Working Group, See also: http://www.w3.org/2001/sw/WebOnt/, and http://www.ontoknowledge.org/oil/misc.shtml#ackn

Wagner, C., Bottermann, M., Freijen, E., Schmidt, L., Simmons, S., Van ´t Hof, C., et al. (2004). *Europe, competing: Market prospects, business needs and technological trends for virtual smart organisations in Europe*. Report for the European Commission, DG Information Society, RAND Corporation, MG-195-EC.

Warnecke, H. -J. (1992). *Die Fraktale Fabrik. Revolution der Unternehmenskultur* [The fractal company]. Berlin; New York: Springer-Verlag.

Weiser, M., & Brown, J. S. (1998). Center and periphery: Balancing the bias of digital technology. In D. Tapscott, A. Lowy, & D. Ticoll (Eds.), *Blueprint to the digital economy: Creating wealth in the era of e-business* (pp. 317-335). New York: McGraw Hill.

Wickens, P. D. (1998). *The ascendant organization* (revised and updated edition). London: MacMillan.

Wiesenfeld, B. M., Raghuram, S., & Garud, R. (1998). Communication patterns as determinants of organizational identification in a virtual organization. *Journal of Computer-Mediated Communication, 3*(4).

Williamson, O. E. (1975). *Markets and hierarchies: Analysis and anti-trust implications*. New York: The Free Press.

Witczynski, M., & Pawlak, A. (2002). Virtual organizations enabling net-based engineering. In B. Stanford-Smith, E. Chiozza, & M. Edin (Eds.), *Challenges and achievements in e-business and e-work* (pp. 908-915). Amsterdam: IOS Press.

Wognum, N., Thoben, K. -D., & Pawar, K. S. (Eds.). (1999). *ICE'99, International Conference on Concurrent Enterprising*. Nottingham: University of Nottingham..

Zobel, R., & Filos, E. (2002). Work and business in the e-economy: technology & policy issues. In B. Stanford-Smith, E. Chiozza, & M. Edin (Eds.), *Challenges and achievements in e-business and e-work* (pp. 52-66). Amsterdam: IOS Press.

Section II

Technologies for Operation of Smart Organizations

Chapter II

Applications of Agent-Based Technologies in Smart Organizations

László Zsolt Varga, Hungarian Academy of Sciences, Hungary

Abstract

This chapter introduces agent technology as a means of creating dynamic software systems for the changing needs of smart organizations. The notion of agency is introduced, and individual and collective agent architectures are described. Agent interaction methods and agent system design techniques are discussed. Application areas of agent technology are overviewed. The chapter argues that the autonomous and proactive nature of agent systems make them suitable as the new information infrastructure for the networked components of dynamically changing smart organizations.

Introduction

Nowadays the whole world is networked into the Internet and if an organization is not connected to the Internet, then it has serious competitive drawbacks. Private persons are using the Internet more and more as well, so organizations keep contact with their clients through e-mail and give them information on their products and services on information portals. Customers can do the shopping in electronic shops and get all the information they want from the portal server; they can even configure the product they want to order. In order to satisfy individual needs, smart organizations must feed online information from the Internet into their internal information system and then further to their internal production control, accounting, design, resource planning, and several other components. The organization can adapt to these requirements only if it requires the same type of information management from its suppliers, so the interorganizational communication must become part of this networked environment as well.

In this environment, we can less and less talk about individual software products, because software components are interconnected and sooner or later almost every software component must be capable to interoperate with other software systems. This way, the information system of smart organizations becomes part of the worldwide Internet, so individual solutions cannot be applied. The software technology of smart organizations means less and less the design and implementation of individual software systems; rather, we can talk about the development of the design and implementation of a single distributed worldwide information system. In this context, the designers of subsystems cannot apply individual solutions, they have to adapt to global practice and standards. At the time of the design of such a global information system, the designer does not have enough information and resources to make a complete solution, so the designed system must integrate into the worldwide system with the ability to adapt to unforeseen changes and requirements using incomplete information at run-time.

Satisfying these requirements is among the goals of several technologies, including the Web services technology characterized by SOAP[1], WSDL[2], UDDI[3] abbreviations (Web Services, 2004; UDDI, 2004), the semantic Web technology (Berners-Lee, Hendler, & Lassila, 2001), the grid (Foster & Kesselman, 1999) and maybe the most complete approach, which is agent-based computing (Wooldridge, 2002).

This chapter presents the most important elements of agent technology and how they can be applied in smart organizations. First, we define what agents and agent systems are, then we overview the history of agent developments. We discuss the internal structures of agents, then how these agents can form smart organizations, then the methods of agent system analysis and design. Finally, we discuss the applications of agent systems and the conditions of their wide adoption.

The Agent Metaphor

The word "agent" has different meanings in different contexts, so computer scientists working in the agent field may have somewhat different definitions of agency. There is agreement on the main characteristics, but some researchers consider other characteristics important as well, while some researchers think that these are not important, depending on their background.

Intelligent Agents

The notion of agent emerged from many different fields, including economics, game theory, philosophy, logic, ecology, social sciences, computer science, artificial intelligence, and later distributed artificial intelligence. In all these fields, an agent is an active component that behaves intelligently in a complex environment to achieve some kind of goal. Artificial intelligence is the branch of computer science which investigates how to implement in computer systems intelligence comparable to human intelligence. While the goal of artificial intelligence focuses mainly on intelligent performance comparable to an individual person, distributed artificial intelligence investigates how a group of software components called agents can achieve intelligent behavior comparable to a group of persons.

From a software technology point of view, agent technology promises to enable system designers to handle more complex systems than before. As systems become more and more complex, software development processes need higher and higher abstractions. In the beginning, functional and modular programming techniques provided enough level of abstraction, then object-oriented systems became the most commonly used technique to model complex systems. Agent technology promises to handle systems that object-oriented

Figure 1. The most important characteristics of intelligent agents

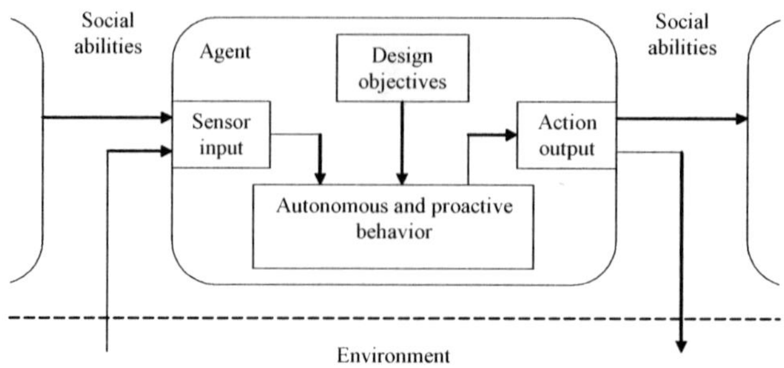

techniques cannot adequately model, like large, distributed organizations with incomplete information and distributed responsibility, where individual components must dynamically adapt to unforeseen changes.

Experts from the different fields tend to agree that the most important characteristics of agents are those which are defined by Wooldridge and Jennings (1995) and shown in Figure 1. First of all, an agent is a *computer system situated in some environment*. The agent is *reactive,* which means that it is capable of sensing its environment and acting on it. The agent can *autonomously act* in its environment and make decisions itself. The agent has *design objectives* and can decide itself how to achieve them. While taking the decisions the agent is not just passive, but can *take initiatives* towards its goals. The agent has *social abilities* and can interact with the actors in its environment.

Agents as Building Blocks in Smart Organizations

The above-mentioned characteristics make the agent concept an important element in modeling systems needed for smart organizations. First of all, multi-agent systems are distributed cooperative computing systems, therefore they themselves form an intelligent organization. The reactive, autonomous, and proactive features of agents require that they are knowledge-driven, dynamically adaptive, agile, and learning computing elements. The social abilities of agents mean that they are usually internetworked, as well as dynamically adaptive to new organizational forms and practices. Since these features are necessary for smart organizations, we can expect that software systems built with agent technology will play an important role in smart organizations. A multi-agent system itself can be regarded as a smart organization, because the above-mentioned characteristics are in line with the definition of smart organizations. The term "smart organization" is used for organizations that are knowledge-driven, internetworked, dynamically adaptive to new organizational forms and practices, learning, as well as agile in their ability to create and exploit the opportunities offered by the new economy (Filos & Banahan, 2000).

In the following sections, we will discuss agent systems in order to be able to understand their importance for smart organizations. Agent technology offers new techniques for smart organizations, but it cannot solve everything. Several design techniques and software tools have been developed to support and implement agent oriented systems. Although these techniques and tools allow the designer to think in the way an agent system needs, the major challenge in implementing agent systems is related to the intelligence of the agents.

History and Standards

Current interest in autonomous agents emerged mainly from artificial intelligence research, but object-oriented programming and human-computer interface design also contributed among the many other fields mentioned earlier. Although we could think that agency is central to artificial intelligence (AI), because AI is about building intelligent systems, artificial intelligence researchers did not intensively study intelligent agents until the 1980s. The focus of AI research was on the different components of intelligent behavior, like learning, reasoning, problem solving, and so forth. Among these independent investigations, AI planning was most closely related to agents, because AI planning is related to what and how to do, and agents also have to plan what they are going to perform autonomously in their environment. AI planning first used a symbolic reasoning approach, but when the ultimate viability of this approach was questioned, the attention of researchers turned toward behavioral or reactive AI. According to this approach, theorem provers cannot produce intelligent behavior; rather, intelligence is a product of the interaction between the intelligent system and its environment. In this approach, intelligence emerges from the interaction of several simpler behaviors and competing behaviors can suppress each other. However, emergence is purely reactive, so in the early 1990s researchers started to combine reactive behavior with the deliberative approach of symbolic reasoning. The combination of reactive and deliberative approaches was later replaced with the practical reasoning approach, where reasoning is influenced by a kind of mental state with three components: Beliefs, desires, and intentions, where beliefs represent the information that the agent has about its environment, desires represent the different possible states the agent may choose to commit to, and intentions represent the states the agent has chosen and committed resources to.

The Beginnings

Agent research became a separate branch of AI in 1980 at the first Distributed Artificial Intelligence (DAI) workshop at the Massachusetts Institute of Technology, where participants decided that there is need to investigate issues of how intelligent problem solvers can coordinate their activities to solve common problems, and these issues are on a higher level than the parallelism issues of how to distribute processing over machines and parallelize centralized algorithms. The first multi-agent model was the actors model, in which self-contained, interactive components communicate by asynchronous message passing. Task allocation then became an important topic, and the Contract Net Protocol was defined to allocate tasks from the contractor to bidders through an announcement—bidding—allocation process. The early applications were related to the coordination of physically moving vehicles. Later, the research focused on teams working toward a common goal, and theoretical foundations of cooperation were investigated, including notions of commitment and joint intention. A group of agents jointly intends a team action if all of them are committed to completing the team action and they mutually believe that they are doing it. In this case, the joint commitment is a joint persistent goal. Agents enter into a joint commitment by establishing appropriate mutual beliefs and commitments through an exchange of request and confirm speech acts.

The investigation of how to achieve joint commitment centered on the notion of negotiation. It turned out that negotiation was a good method for coordination, conflict resolution, communication of plan changes, task allocation, and resolution of constraints violations as well. The common characteristics of these are that agents have to resolve some conflict in a distributed way by exchanging proposals and counter proposals, the agents have their own goals, they have bounded rationality and incomplete information.

At this time, agent architectures focused on the internal modules of agents and how the above-mentioned concepts can be handled with software engineering methods. Agents usually had five components: the communication layer, the agent acquaintance module, the self module, the inference engine, and the knowledge base. The communication layer was responsible for performing the necessary transformations on the messages the agent wanted to send and receive to and from its environment, in order that these messages conform to the external and the internal world of the agent. The agent acquaintance module contained information about the environment of the agent and modeled the capabilities of the agents to interact with. The self module contained information about the capabilities of the agent itself. The inference engine was responsible for executing the actions of the agent based on the knowledge of the agent stored in the knowledge base.

Networked Agents

In the 1990s, the Internet and hypertext protocol was spreading rapidly, and more and more applications were deployed on the Internet. This open environment gave way to the wide-spread application of software agents communicating over Internet protocols. Previously, multi-agent systems were designed and implemented usually by a single team, but now multi-agent systems from different backgrounds and design approaches had to communicate and interact. The most important issues in this environment are discovery and interoperability. Discovery is the problem of how agents can find each other even when they do not know anything about the other agent. Interoperability is the problem of understanding the syntax and semantics of the language of other agents, which means that agents have to be able to parse the message of other agents and find out the meaning of the elements of the messages.

To solve the discovery issue in open environments, the notion of middle agents has been introduced. Agents can advertise their capabilities to some kind of middle agent. Different types of middle agents have been identified, including yellow page middle agents that match advertisements with requested capabilities, blackboard middle agents that collect requests, and broker middle agents that do both. Of course, this middle agent approach works only if agents know how to find the appropriate middle agent. In practice, this can be solved by having a few well-known middle agents, which preferably even talk with each other so that if there is no match at a specific middle agent, then the request can be forwarded to another one. These well-known middle agents form the basis of the infrastructure of an open agent environment. The first attempt for a world wide agent infrastructure was the Agentcities network (Willmott, Dale, Burg, Charlton, & O'Brien, 2001).

Solving agent interoperability is approached on two levels: on the agent communication language level and the agent content language level. The agent communication language defines the types and the format of the messages between agents. Agent communication languages provide a set of performatives, like "request" and "inform," based on the speech acts theory (Searle, 1969), where communications are modeled as actions that change the mental state of communication participants. Using the agent communication language, an agent can send to another one a request for "something," or can inform the other agent about "something," where the "something" is the content of the message. The schema for the agent content language is the ontology which formally describes a domain of discourse. Agents can understand the content of the messages if they share their content language ontologies, preferably by publishing them on ontology servers. Ontology servers are also an important part of an open agent infrastructure like the Agentcities network (Willmott et al., 2001)

Standards and FIPA

The need for interoperable agent communication created the standardization body of agent systems, which is called Foundation for Intelligent Physical Agents (FIPA). FIPA was founded in 1996 and registered in Geneva, Switzerland as an international nonprofit organization. The aim of FIPA is to develop software standards for heterogeneous and interoperating agents and agent systems, in order to enable the interworking of agents and agent systems operating on platforms of different vendors in industrial and commercial environments. As a result of the FIPA standardization activity, many research labs and industrial organizations started to develop competing agent platforms independently all over the world. FIPA standard agent platforms provide an environment where agents can be deployed, and with the help of the agent platform services they can interact with other agents on any FIPA standard agent platform in a FIPA conformant way, achieving agent communication level interoperability. Agent platforms from more than 15 vendors show interoperability in the Agentcities testbed. More than half of the Agentcities nodes use the Jade agent platform from Telecom Italia Laboratories (Balboni, 2003).

The most important agent standardization activities are done in FIPA, but significant activity was also carried out in the Object Management Group

Figure 2. Trends in agent research

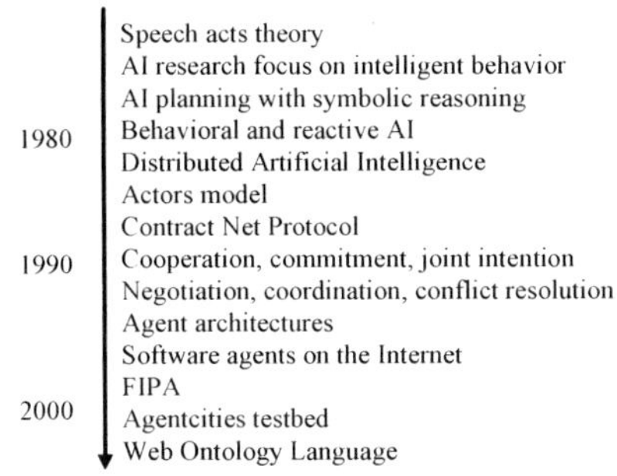

(OMG) and agent standards are starting to become highly relevant to bodies such as the World Wide Web Consortium (W3C) and the Global Grid Forum (GGF), because developments such as Web Services (Web Services, 2004) and Semantic Web Services (DAML Services Coalition, 2002; Bussler, Maedche, & Fensel, 2002) also investigate many of the issues agent technologies have already addressed.

Figure 2 summarizes the history and trends in agent research as discussed in this section.

Agent Architectures

As we have seen in the previous section, the agent concept evolved over time. Different aspects of agency were discovered and in the end merged into the currently applied agent architectures. Nowadays, agents that show traits of only one aspect are not considered real intelligent agents. For example, a stock exchange trading agent in charge of a stop-loss order is a purely reactive agent, but does not satisfy the current notion of intelligent agency. In this section, we are going to elaborate on the different aspects of agency.

Reactive Agents and Agents with State

One of the first aspects is that agents are reactive. A *purely reactive agent* decides what to do without reference to its history. The behavior of a purely reactive agent is the function of the state of its environment. This type of agent architecture has two main subsystems: perception and action. The perception subsystem contains the agent's ability to observe its environment. In the case of agents in the physical world, like robots, this may be a video camera, and in the case of an agent in the software world this may be system or network routines like finger, ping, or network messages. The output of the perception module is a percept, or the internal representation of the environment. The action subsystem of the agent contains the agent's ability to act on its environment. In the case of a physical agent this may be a robot arm, and in the case of a software agent this may be system commands. The action subsystem maps the sequences of percepts into actions. The perception subsystem of the agent grabs those features of the environment which are relevant for the goals of the agent. For example, in the case of the stock exchange trading agent in charge of a stop-loss order, the whole range of the stock price is mapped into two values: hold and sell. If the price falls below a certain value, then the agent has to issue a sell order.

Purely reactive agents often compose a *fine-grained* multi-agent system. A fine-grained multi-agent system consists of many simple agents, and the intelligent behavior of the fine-grained multi-agent system emerges from the interaction of the simple agents. *Coarse-grained* multi-agent systems consist of fewer, but more intelligent, agents. Agents in a coarse-grained multi-agent system usually have one of the architectures discussed below.

Purely reactive agents do not remember the history of their environment. *Agents with state*, shown in Figure 3, can do so by having additional components in their architecture: a state and a next function. The state represents the current mental status of the agent, while the next function maps the percept of the agent and the current state of the agent to the next state of the agent. The action subsystem of agents with state maps the current state of the agent into actions. Agents with state have the full power of agency; they are

Figure 3. Agent with state

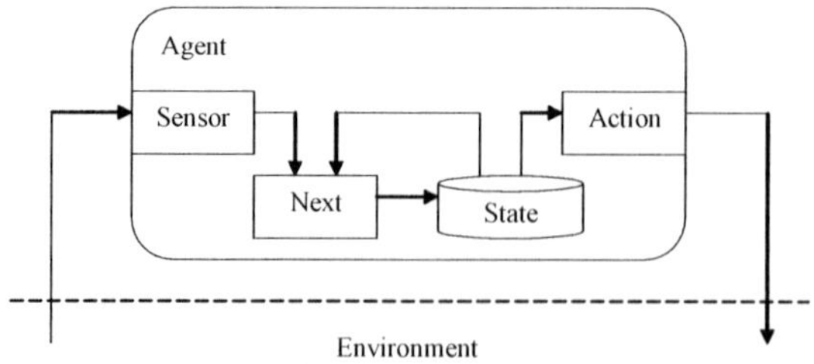

behaviorally equivalent to agent architectures discussed later in this section, but the other architectures grab more of intelligent behavior and help better understand the notion of agency.

Agent Reasoning

Agents usually have to act in a dynamically changing environment, therefore it is better not to tell the agent exactly how to carry out the tasks. It is better to tell the agent what to do without telling how to do it. This can be done by defining tasks indirectly with some kind of performance measure.

One way of defining tasks indirectly is by associating utilities with states of the environment. The *utility function* maps the environment states to real numbers and tells how good the state is: the higher the value, the better the state. A disadvantage of the utility function is that it assigns utilities to local states and does not take into account long term values. However, we can use overall utility; for example, by defining the worst state that might be encountered by the agent or as the average utility of all states encountered. Although this model is useful to understanding agent behavior, in practice sometimes it is very difficult or even impossible to implement the desired utility function.

Another way of indirect task specification is *predicate task specification*. Predicates are utility functions that have either true or false values. A predicate task specification maps the set of all possible runs of the agent to true or false value, and the agent achieves the desired goal if the predicate function results in true value either for all runs, or at least for one run or for a given percentage of runs of the agent, depending on how pessimistic or optimistic the definition of success is. Some common forms of predicate task specifications are the achievement tasks and the maintenance tasks. In the achievement task the goal of the agent is to achieve a state, while in the maintenance task the goal of the agent is to maintain a state. In the achievement task, the agent is successful if it can force its environment into one of the goal states, while in the maintenance task the goal of the agent can be characterized as to avoid some state. Complex tasks can be specified as combinations of achievement and maintenance tasks.

Deductive reasoning agents originate from symbolic AI, which says that intelligent behavior can be generated using logical deduction or theorem proving from symbolic representation of the world. In this approach there are two key problems: the transduction and the reasoning problem. The transduction problem is how to translate the real world into an accurate and adequate

Figure 4. Deductive reasoning agent

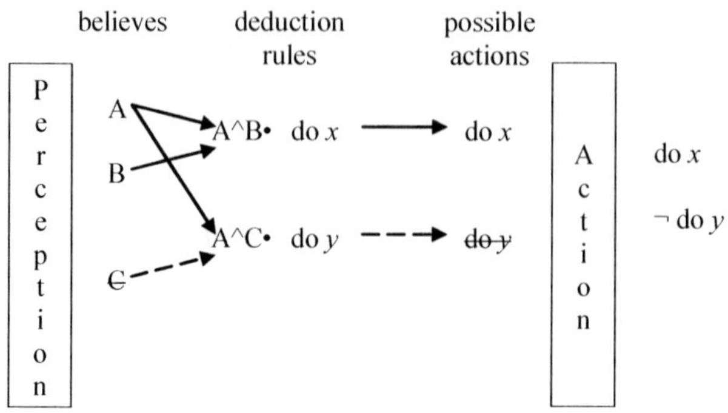

symbolic representation. This may be very hard; for example, in the case when a photo has to be converted into a set of declarative statements representing that photo. The reasoning problem is how to manipulate symbolic information to be useful in time. Since the computational complexity of theorem proving may require long computation, theorem provers may not always operate effectively in time-constrained environments. A deductive reasoning agent (shown in Figure 4) is an agent with state and its perception module translates external information into symbolic representation. Once there is a symbolic representation of a fact in the database of an agent, then the agent believes this fact, although in the real world this might not be the case. The next function of the deductive reasoning agent maps the agent database and a perception into a new database. The action subsystem of deductive reasoning agents use deductive reasoning to deduce the action of the agent. The deduction rules of the agent are defined in a way that if a formula "do action A" can be derived from the fact database using the rule database, then the action of the agent will be action "A." The reasoning engine of the action subsystem takes each of the possible actions "x" of the agent and tries to prove "do action x." If there is no action for which this formula can be proved, then the reasoning engine tries to find an action "x" for which "do not do action x" cannot be derived. If there is such action "x," then this action is consistent with the rules, so the agent can execute this action. If this also fails, then the agent does nothing.

Practical reasoning agents try to improve the deductive reasoning agent architecture by reducing the search space of deductive reasoning. One of the main problems of deductive reasoning agents is that deducing all possible logical consequences takes too much time and sometimes is even impossible.

In practical human reasoning, the logical reasoning is influenced by the current state of the mind. Human practical reasoning first tries to reduce the search space by deciding what state we want to achieve. This is called *deliberation*. Once deliberation is done, the reasoning concentrates on how to achieve the selected state. This is called *means-ends reasoning*. There must be a good balance between deliberation and means-ends reasoning, or else practical reasoning agents do not perform well or even do nothing. Deliberation cannot go on forever—some goal state has to be chosen and the process of achieving this state has to be started even if the selected goal state is not optimal. The process of achieving the selected goal state is called *intention*. Intention involves the process of creating a plan to achieve the selected goal state and actions taken according to the created plan. Deliberation and intention are shown in Figure 5. Intentions drive the means-end reasoning and if one plan creation fails, then another is tried. An intention must persist typically until it is believed that it is successfully achieved, or it is believed that it cannot be achieved, or it is believed that the reason for the intention no longer exists. Intentions constrain deliberation, and options which are not consistent with the current intentions are dropped. Intentions restrict the beliefs on which practical reasoning is based, and beliefs that are not consistent with the intention are dropped.

The deliberation process of practical reasoning agents has two parts: the option generation function and the filtering function. The option generation function produces a set of options, called *desires*. The filter function selects the best one(s) from the set of desires based on the current beliefs, desires, and intentions.

Once a desire passes the filter function and becomes part of the set of currently selected intentions of the agent, then we say that the agent has made a *commitment* to that intention. The commitment strategy of the agent is the mechanism used to determine how long a commitment must persist. Blind commitment strategy is to keep the intention as long as the agent believes that the intention has been achieved. Single-minded commitment strategy is to keep

Figure 5. Steps of practical reasoning

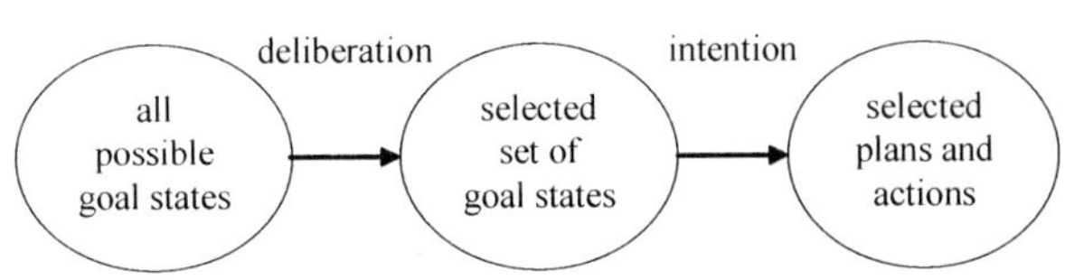

the intention as long as the agent believes that either the intention has been achieved, or the agent currently has no plans to achieve the goals of the intention. Open-minded commitment strategy is to keep the intention as long as the agent believes that the goals of the intention are possible. The agent has to reconsider its commitments from time to time to check if they still have to be kept. There must be a good balance, because if the agent reconsiders its commitments very often, then it does not have enough computing resources to achieve them; on the other hand, if the agent does not reconsider its commitments often enough, then it may continue to pursue them for a long time after it is obvious that they cannot be achieved.

Means-ends reasoning produces a plan to achieve the selected goal state based on the current intentions, the current beliefs (i.e., the state of the environment), and the actions available to the agent. In many implementations, the planning function does not create a plan from scratch; rather, the agent has a set of plans given by the agent designer, and the agent searches through the set of plans to find one that has the needed intention as a post condition and is in accordance with the current beliefs and available actions.

Layered Agent Architectures

An alternative to the reasoning agent architecture is the *hybrid agent architecture,* or *layered agent architecture,* in which there are layers responsible for different agent-like behaviors. In the horizontally layered hybrid architecture, each layer is directly connected to the perception and the activation modules, as shown in Figure 6. In horizontal layering, each layer produces competing suggestions as to what to do, and a control subsystem must decide which layer actually takes control over the agent. Some of the layers are responsible for low-level actions; for example, in a financial organization, to avoid bankruptcy some of the other layers are responsible for higher-level actions like deciding

Figure 6. Horizontally layered agent architecture

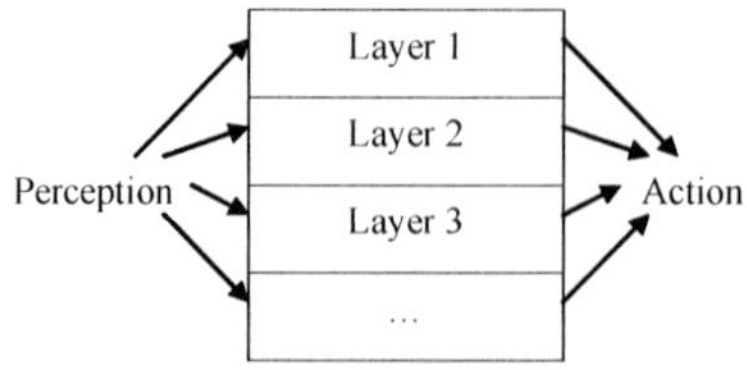

Figure 7. Vertically layered agent architectures

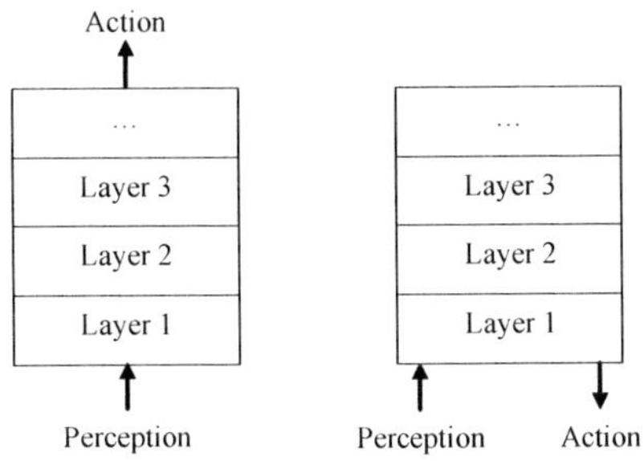

where to invest. The control subsystem gives priority to low-level actions in urgency and gives way to higher levels otherwise.

In vertically layered hybrid architectures, perception and activation are connected to at most one layer, as shown in Figure 7. Layers make processing and pass information to each other. In two pass vertical layering information flaws up the architecture to higher and higher level processing. Decision is made at the upper-most level where action is generated, which then flaws down to lower levels.

Not long ago, a popular agent model was the *mobile agent* architecture. In this model, agents are seen as programs roaming the network to collect business-related data. This approach had a lot of problems related to authorization policies; that is, hosts and agents had to be protected against each other. Since network bandwidth is usually available, mobile agents did not have much advantage over nonmobile agents except in a few cases—for example, in auctions when different network latencies were not allowed for fairness. Because of the difficulties, mobile agents have not yet been taken up by the mainstream; however, mobility issues may be investigated again when agents running on mobile devices become widespread.

Agent Organizations

Up to now, we have been discussing how agents can organize and plan their activity on their own, but agents have to act in a networked environment; for example, as part of a smart organization. In this environment, they act on the real world, and sometimes the real world imposes restrictions on their activities; for example, because two agents want to use the same resource at the same time. In this case, it is obvious that the agents have to coordinate their plans. Even if there is no conflict in the real world, the agents want to distribute the task allocations among themselves, and there must be some kind of interaction between the agents. The interactions are even more complex when the conflict in the real world arises between the activities of two groups of agents. In this case, the groups of agents have to coordinate among themselves as how to interact with the other group.

In order to model the interactions among agents, the *utility function* is used. The utility function of an agent assigns to each state a real value. If the utility function gives higher value for a state s_1 than for another state s_2, then the agent has *preference* for s_1 compared to s_2. Many times the utility function is linear, but nonlinear utility function models that situation when the agent achieved most of its goals and is satisfied with the state, therefore its utility function does not give much higher values when the state improves somewhat. Similarly, if the agent has not achieved any goal, then a small improvement in the state gives higher increase in its utility function than in a more or less satisfied status.

Properties of Agent Organizations

When several agents act on the environment, their actions may depend on the actions of the other agents. If one agent makes a choice, then the other agent is already restricted and has to make a choice depending on the choice of the other agent. In an ideal situation, the different agents have preference for the same state and all other states are less preferable for all of the agents. A somewhat less ideal but still very good situation is when agents can still find a state which is most preferable for all the agents, but there are other states which give the same utility value for all the agents. In this case, agents can select one of these preferable states, but they must agree which one, because if an agent deviates from this state toward another more preferable state, then none of the agents achieve the most preferred state. It is also possible that there are more than one state with which agents are all satisfied and do not want to deviate from it if the others do not deviate; however, one of these preferable states may be better than the other one. All the situations in this paragraph are called *Nash equilibrium*, because no agent has the incentive to unilaterally deviate from the preferable state.

The *efficiency* of the agent system can be measured as a combination of the utility functions of all of the agents. A simple efficiency measure is the sum of the utilities of all the agents, and according to this measure an agent system is in *sum optimal* state if the sum of the utilities is maximal. An agent system is in a *Hicks optimal* state if the utility is maximized for all of the agents in the agent system. An agent system is in *Pareto optimal* state if it satisfies, more or less, all of the agents, and in all other states at least one agent's utility function gives smaller value if at least another agent's utility function gives higher value. Note that Hicks optimal state cannot always be achieved. Also note that sum optimal and Pareto optimal state may not be equilibrium state, if at least one agent might achieve better utility by deviating from the optimal state. An example of this is the prisoner's dilemma, in which the equilibrium is not optimal.

Agreement in Agent Organizations

Now that we have seen the different types of states multiple agents can achieve, let us turn our attention to how they can reach agreement to get to the desired state. Agents coordinate their actions by exchanging messages. The messages are exchanged similarly to usual network communication protocols, which are governed by protocol rules so that the participating partners can get to some useful result and are not locked in, for example, a deadlock. Agent interaction protocols build on communication protocols and strive to ensure, for example, community level results (Sandholm, 1999). It is expected that an agent interaction protocol guarantees that agents eventually get to some agreement and this agreement leads to either sum, Hicks, or Pareto optimal state. Participating in agent interaction protocols must be Nash equilibrium behavior for the participating agents, that is, all of the agents must be interested in keeping to the protocol rules, which must be simple enough so that agents can easily determine the optimal strategy. Multi-agent systems are usually distributed and there is no centralized node, and this must be the case for agent interaction protocols as well.

Although agents may interact in many different ways, there are three types of interaction protocols which are the most used and studied. These are the auction, the negotiation, and the argumentation interaction protocols.

The *auction* protocol can be used to allocate a given resource to one of the agents from a group. The resource can be a good or a task to be executed; in the latter case the auction protocol is also called a contract net protocol. The roles in the auction protocol are the auctioneer and the bidder. The auctioneer agent has the resource to be allocated and wants to maximize the price for it. The bidders are the agents to which the resource is to be allocated and want to minimize its price. In many cases the exact value of the resource is not known or is not unambiguous. The agents may value the price of the resource differently according to their different interests in the resource and different knowledge about the current and future value of the resource. The auction protocol helps the agents agree on a price and allocation which is most acceptable for them. According to the different rules, the auction interaction protocol can be one shot, if there is only a single round of bids, or it can be ascending or descending sequential, if there are several rounds with the necessity of ascending or descending bids. The auction interaction protocol can be open cry, if every agent sees the bid of every other agent, or can be sealed bid, if they do not see each other's bid. The auction protocol is first price if the winner is the one with the best bid and pays its own bid, or it can be second price if the winner is the one with the best bid, but pays the second best bid.

Auction is a special form of *negotiation* which is a somewhat more general form of agent interaction. The negotiation interaction protocol is defined with the negotiation set, the proposal order, a set of strategies, and an agreement criterion. The negotiation set contains all the possible proposals the agents can make. In the simplest case, the proposal contains one issue to be negotiated, like the price in auction protocols, or can contain multiple issues which may be interrelated. The proposal order defines the set of allowed proposals as a function of the negotiation history and the timing of proposal making. Typically, agents make the proposals at the same time or one after the other, and they are not allowed to repeat previous proposals. Each agent has a negotiation strategy which defines the proposal selection method from its allowed proposals. Negotiation strategies are not public and are related to how the agent is going to achieve its goal. The agreement criterion defines when the negotiation stops and what the accepted proposal is.

The most complex form of agent interaction protocol is *argumentation,* which most resembles human negotiation and allows dynamic negotiation and the justification of the negotiated deal. Argumentation is based on formal logic. In formal logic there are statements and logic rules. Using the logic rules other statements can be derived from a set of statements. In the beginning of the argumentation interaction protocol, the agents have in their knowledge base different sets of statements which represent their beliefs about the state of affairs. During the argumentation process agents can send each other the statements they have, the derivation rules they are capable of, and concrete derivation instances in order to get to a status when all the agents have the same statement about the issue to be agreed in their knowledge base.

Communication in Agent Organizations

We have seen how agents can get to agreements by exchanging messages; now let us see how they communicate these messages. In usual-distributed computing environments, like in object-oriented systems, one object can call the method of another remote object. In this kind of communication, the calling object causes the execution of the procedure in the remote object. This may happen synchronously if the thread of control returns to the calling object only after the execution of the remote procedure, or asynchronously if the thread of control immediately returns to the calling object and the remote procedure is executed in parallel. In both cases the calling object executes the remote procedure. However, agents are autonomous and their communication is even less coupled, as in the asynchronous remote procedure call.

Agents are autonomous, and when agents send messages to each other, they do not force the execution of a remote procedure, or write data into the internal data representation of the remote agent. When a sending agent sends a message to a remote agent, the sending agent performs an action to influence the behavior or the beliefs of the remote agent. This kind of communication is based on the *speech act theory,* which treats communication as action (Austin, 1962; Searle, 1969). When a fact is sent from the sending agent as an information to the remote agent, the sending agent intends the remote agent to believe the fact, but it is up to the remote agent whether it trusts the sending agent and builds the belief into its knowledge base or not, as shown in Figure 8.

Figure 8. Autonomous communication

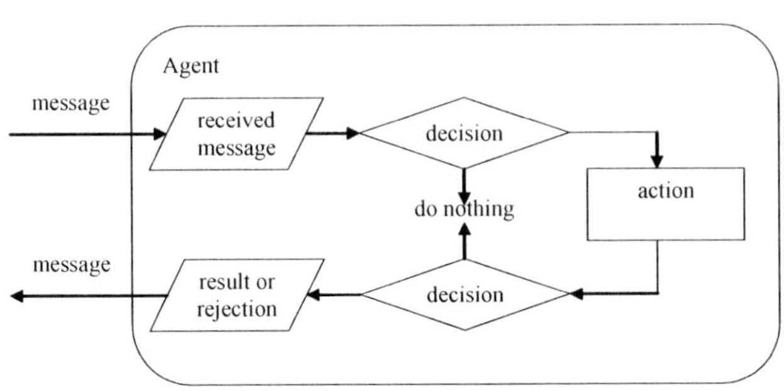

Based on the speech act theory, the Knowledge Query and Manipulation Language (KQML) was developed (Finin, McKay, Fritzson, & McEntire, 1993) in the framework of the DARPA funded Knowledge Sharing Effort. The KQML language defines the envelope format for agent messages, and the content of the message is described in the Knowledge Interchange Format (KIF) (Genesereth & Fikes, 1992). The KQML envelope contains what the intention of the sending agent is with the information contained in the content part. The KQML part of the message has slots for the type of the message (inform, request, reply, etc.), the sender, the receiver, the language of the content (e.g., KIF), the ontology the content is related to (e.g., electronic products), the content itself, and possibly other features.

We should say a few words about the content language and the ontology. The content language is the format of the description of the content. However, the content cannot be anything, it belongs to a specific domain of discourse that both agents understand. The *ontology* specifies the notions of the allowed content, the possible properties of the notions, and relations between the notions of the domain. Roughly we could say that the content is the data and the ontology is the schema of the content, but the ontology defines not only the syntax of the allowed content, but the semantic dependencies as well. An ontology describes the common understanding of a specific domain of discourse; it is described in an ontology description language, and it is usually published so that everybody can use it to understand the same. KIF itself is an ontology language, but the most recent ontology language used on the Web is the Web Ontology Language (OWL) defined and standardized by the World Wide Web Consortium (Dean & Schreiber, 2004).

Although KQML defined a framework for agent communication, it was never precisely defined, therefore many versions of KQML were implemented and when agents started to inhabit the Internet, they could not interoperate. Based on the KQML efforts, FIPA standardized agent communication with the specification of the FIPA Agent Communication Language (ACL), interaction protocols, and architecture. As a result of the standardization effort, many vendors implemented agent platforms interoperable on the communication language level.

Trust and Security

In industrial and business environments, special attention has to be made to trust and security aspects, especially in the open and dynamically changing society of agents forming smart organizations. In the open and dynamic environment, agents interact with each other on an occasional basis without having reliable information on each other and the organization they represent. As identified by Wong and Sycara (1999), the most important security threats in agent systems are the corrupted naming and directory services, the insecure communication channels, the insecure delegation, and the lack of accountability.

A naming service in a distributed environment maps names of components to their addresses. A directory service maps services and capabilities to their providers. These services are not part of agent architectures; rather, they are part of the infrastructure of an agent society. However, the agent society cannot function if the members cannot find each other and their services. A naming service or a directory service is corrupted if some entries are missing or contain a wrong value. A wrong value may be entered, for example, by a misbehaving agent.

The communication channels are secure if authentication, integrity, confidentiality, and nonrepudiation are guaranteed. Authentication means that agents know that they talk to agents they think they are talking to. Integrity of messages is guaranteed if the message is not modified or falsely inserted in the communication channel. The message is confidential if other agents cannot intercept the message. Nonrepudiation is guaranteed if nobody can deny having sent a message which was sent.

Insecure delegation occurs if an agent impersonates itself as a delegate of someone who did not entrust to it. Lack of accountability occurs if agents cannot be held accountable for what they are doing and their services cannot be trusted.

As proposed by Wong and Sycara (1999), several measures have to be taken in an agent society to give protection against the above-mentioned trust and security threats. The naming and directory services must service only valid requests coming from a rightful requester and the request is valid. The naming and directory service databases must be kept consistent. The agents and their delegators must have unique identity which can safely be proved. The communication channels must be protected. Agents can be deployed only if there is someone who can be made liable for their actions.

Theoretical models have been developed to guarantee these protective elements, and agent architectures implement more and more of them.

Agent-Oriented Software Engineering

In the previous sections, we wrote about how agent systems work, but if we want to write about how such agent systems can be designed and implemented, then we have to select from several approaches. Many agent systems have been implemented and now there are dominant standards, tools, and platforms to operate them, but there is no unique methodology for their design. As we have seen, agent systems propose solutions to problems which the traditional software products do not cover; therefore, we cannot expect that traditional software engineering techniques provide solution to agent-oriented software engineering. Of course, when a specific component of an agent system has to be implemented, then traditional software engineering methods can be applied and traditional software components can be used in the implementation. However, we need new methodology until we get to the point at which we can apply traditional techniques.

Usually an analysis and design method provides techniques to understand the problem domain and how to handle the complexity of the system so that it can be designed. This is usually done by creating models of the system at different levels, and then transforming higher level models to models closer to the implementation using formal guidelines. What are needed for agent-oriented software engineering are those high-level models that are above the traditional software engineering methods.

There are many agent-oriented software engineering approaches (Giorgini, Müller, & Odell, 2002) among which probably the agent extensions of UML (Odell, Parunak, & Bauer, 2001) and the Gaia methods (Wooldridge, Jennings, & Kinny, 2000) are the most well known. We are going to write about the latter one here, because this method focuses mainly on the agent levels of agent-oriented software engineering and is based on the organizational view of the system, which is important for smart organizations.

The Gaia methodology starts from the *requirements statements,* which are the textual and formal descriptions of what the system is supposed to do. The requirements capture phase is independent of the paradigm used for the analysis and design, so traditional methods can be used. The Gaia methodology uses the roles and the interactions models for analysis, and the agent, services, and acquaintance models for design. These models and their dependency, as shown in Figure 9, are discussed in the following paragraphs.

Figure 9. Gaia models

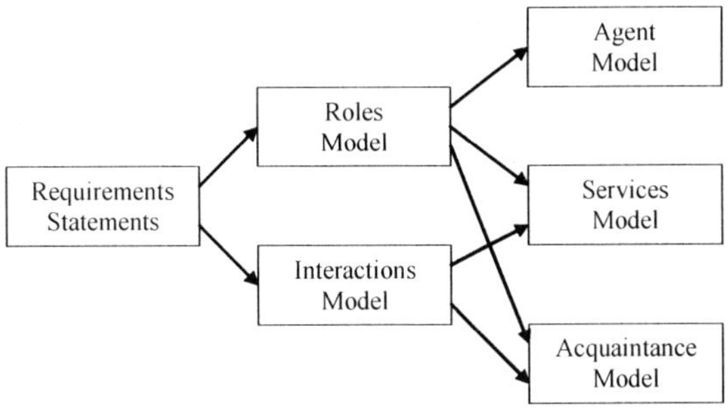

The *roles model* identifies the key roles in the system. The role is an abstract description of the expected function of an entity. The roles are similar to offices in organizations. The role is characterized by the responsibilities and the permissions of the role. The responsibilities are the functions to be performed by the role. A responsibility can either be liveness or safety responsibility. A liveness responsibility says what the role is supposed to do, while safety responsibility is an invariant that the role must keep. Invariants are described as predicates. The permissions associated with the role either identify the resources that can be used to carry out the responsibilities of the role, or define the resource limits within which the execution of the responsibilities can be carried out.

The *interaction model* captures the interaction links between the various roles in a smart organization. The interaction model consists of a set of protocol definitions for each type of interrole interaction. In this model, a protocol is abstracted away from the concrete execution steps and is described by a brief textual description of the protocol, the roles responsible for starting the interaction, the roles with which the initiator interacts, the information used by the initiator while executing the protocol, the information supplied by the initiator and responder roles, and a brief textual description of the processing activities during the protocol execution.

The *agent model* consists of a set of agent types used in the system under development. The agent types are marked with information on how many instances of them will be implemented in the actual system (zero, one or more, n to m, etc.). The agent type is defined as a set of agent roles to be fulfilled by an agent of the given type. Agent types are organized into an agent type tree, where the leaf nodes correspond to roles and the upper-level nodes correspond to agent types. An agent type is composed of the roles of its children agent types in the tree. The agent type tree is derived from the roles model.

The *service model* specifies the functions associated with each agent role. A service is a single coherent block of activity to be carried out by the given agent type. A service is specified with the inputs, outputs, pre-conditions, and post-conditions of the service. The inputs and outputs are derived from the protocol definitions of the interaction model, while the pre- and post-conditions are derived from the safety responsibilities of the roles model.

The *acquaintance model* defines the communication links between agent types. The agent acquaintance model is a directed graph in which each graph node corresponds to agent types and arcs correspond to communication links. Arcs are directed and indicate that an agent of one type will send messages to an agent of the other type. The acquaintance model does not specify what messages are sent or when messages are sent, the goal of the acquaintance model is to identify potential communication bottlenecks.

Using these models the designer can specify most of the agent features of the system under development. Further design and implementation can use any traditional design techniques to implement the agent instances.

Agent Applications

In the previous sections, we discussed what agents are, what their internal structure can be, how agents behave in organizations, and how agent systems can be designed. Now we are going to discuss the applications of agent systems in smart organizations. Basically, agent applications in smart organizations can be classified into three categories: distributed agent systems (or multi-agent systems), assistant agents, and multi-agent simulation systems. In the first two types of applications, agents become part of the smart organization, while in the third type of applications agents are used to evaluate and design the structure of the organization. In the first type of applications, several agents make collective decisions and actions within the organization to improve the operation of the organization. In the second type of applications there may be several agents deployed within the organization, and these agents may even interact with each other, but the main function of each agent is to assist its individual user in autonomous and proactive decision-making.

In *distributed agent system* applications the agent system becomes an integral part of the organization and agents assist the distributed intelligent operation of the organization. Typical areas where distributed agent systems can be applied are business process management, distributed sensing, distributed resource management, process control, trading and purchasing networks. Distributed agent systems can outperform centralized business process management systems, because they are more responsive and are able to cope with unpredictable events. In an agent-based business process management system, the organizational structure and the roles in the organization are mapped to agents, which are responsible for the given role and embody the knowledge needed for the role. These agents can then autonomously and proactively execute most of the automatic processes of the organization with minimal user intervention and approval. Distributed agent systems help distributed sensing by allowing cooperation between the sensors and predicting future trends in the area of one sensor from the data of another sensor. Distributed resource management can benefit from the proactive behavior of agent systems. For example, agents can monitor the network load in telecommunication networks and jointly make predictions on trends and future needs to reallocate resources. Agents can coordinate the workload and the schedule of the field engineers—for example, of electricity provider or telecommunication companies—by taking into account the location and capabilities of the field engineers. Agents can execute the job of automatically negotiating and trading with the suppliers of an organization. Since agents are dynamic, they can adapt to the changing needs of virtual organizations and supply chains.

Assistant agents help their users in gathering and filtering information, or executing some task on behalf of their users. Information retrieval agents can gather information and categorize it according to predefined conditions. This helps the user overview huge amounts of information. More advanced assistant agents learn from the activities of their users; for example, by recording the activities and decisions of the user and deriving rules with knowledge discovery and data mining techniques.

Organizations can also benefit from *multi-agent simulation* systems, which can simulate real-world environments with a high degree of complexity and dynamism. In a multi-agent simulation system, many individual behaviors can be encoded, thus giving a more complex and real picture of what might happen. The organization can make decisions regarding future products and product features based on a multi-agent simulation of the market where the product is to be sold.

Agents Supporting Smart Organizations

As we can see, agent technology discussed in this section has a lot of features that support smart organizations. Smart organizations act in a globally distributed system in which software applications must appear in a new way. A software application in this distributed system is just a component with possible utilizations not completely known at design time. The designer implements some functionalities into the component, but the component may be dynamically included in different temporary compositions in the globally distributed environment. The software component provides services to other software components and it may invoke services of other components. This architecture is in line with the dynamically changing organizations of the economy. Software components advertise their services and other software components can search for the desired services. In order to achieve the goals, software components can select and invoke the desired software services based on the service descriptions, the trust and reputation information available from different sources. In this environment, software components are formed into temporary alliances and their services are dynamically combined. The experiences learned in one temporary alliance are reused in another composition dynamically created later. This way, any software component available on the Web may become part of a Web application.

Agent-based computing provides a new software technology for this new changing environment of smart organizations. Agent technology allows that the creation of the complete functionality of the software system can be postponed beyond design, implementation, and deployment time to operation time, when the software components themselves compose their relation to other software components. This new way of software composition requires that the software components have dynamic and autonomous features.

It is also important that agent technology standards provide the glue for tightening the software components together. Agent technology standards provide machine processable, formal descriptions for the functionality, accessibility, and quality properties of the software components, the data used by the software components, as well as how they can be composed in a workflow. Agent technology also provides standards for registering and searching agents and their services in registries.

Agent technology also takes software components considerably closer to semantic interoperability, which is crucial to smart organizations. Semantics is the relation between the formal notation systems used by the computers and the real objects and notions used by humans. Although simple bit sequences may have semantic meaning, it is better to have the computer representation closer to the human representation, because this way computer interaction on a higher semantic level can be implemented more easily.

Summary

In this chapter, we discussed how agent-based technologies can contribute to smart organizations. Agent technology forms the base of knowledge-driven, internetworked, dynamically changing systems like smart organizations. The most important characteristics of agents are that they are reactive, autonomous, have design objectives, can take initiatives towards their goals, have social abilities, and can interact with the actors in their environment. Agent technology emerged from artificial intelligence by dealing with distributed aspects, and lead to the semantic interoperability technologies of the current Internet. Agent architectures provide means for agents to organize and plan their activity on their own. The types of states multiple agents can achieve can be classified from stability, efficiency, and optimality aspects. Agents reach agreement to get to the desired state by exchanging messages. Agent interaction protocols build on communication protocols and strive to ensure community level results. Agent technology can also be viewed as a software engineering approach to design large, open, networked, dynamic software systems. Agent technology applications can be classified into three categories: distributed agent systems (or multi-agent systems), assistant agents, and multi-agent simulation systems. The methods and approaches discussed in this chapter show that agent technology is fundamental to smart organizations.

References

Austin, J. L. (1962). *How to do things with words*. Oxford: Oxford University Press.

Balboni, G. P. (Ed.). (2003). EXP - in search of innovation [Special Issue]. *JADE*, *3*(3), 1-141.

Berners-Lee, T., Hendler, J., & Lassila, O. (2001). The Semantic Web. *Scientific American 284*(5), 34-43.

Bussler, C., Maedche, A., & Fensel, D. (2002). A conceptual architecture for Semantic Web enabled Web services. *ACM Special Interest Group on Management of Data, 31*(4).

DAML Services Coalition (2002). (alphabetically Ankolenkar, A., Burstein, M., Hobbs, J.R., Lassila, O., Martin, D.L., McDermott, D., McIlraith, S.A., Narayanan, S., Paolucci, M., Payne, T.R., Sycara, K.) *DAML-S: Web Service Description for the Semantic Web.* The First International Semantic Web Conference (ISWC). Sardinia (Italy).

Dean, M., & Schreiber, G. (Eds.). (2004). *OWL Web Ontology Language Reference.* Retrieved February 10, 2004, from http://www.w3.org/TR/owl-ref/

Filos, E., & Banahan, E. (2000). Will the organisation disappear? The challenges of the new economy and future perspectives. In L. M. Camarinha-Matos, H. Afsarmanesh, & R. J. Rabelo, R.J. (Eds.), *E-business & virtual enterprises* (pp. 3-20). Dordrecht: Kluwer.

Finin, T., McKay, D., Fritzson, R., & McEntire, R. (1993, December). KQML: an information and knowledge exchange protocol. *Proceedings of the 1993 International Conference on Building and Sharing of Very Large-Scale Knowledge Bases,* Tokyo, Japan.

FIPA (2001). The Foundation for Intelligent Physical Agents. http://www.fipa.org/.

Foster, I., & Kesselman, C. (Eds.). (1999). *The grid: Blueprint for a new computing infrastructure*. San Francisco: Morgan Kaufmann.

Genesereth, M. R., & Fikes, R. E. (1992). *Knowledge interchange format, Version 3.0 reference manual* (Tech. Rep. logic-92-1). Stanford, CA: Stanford University, Computer Science Department.

Giorgini, P., Müller, J. P., & Odell, J. (Eds.). (2002). Agent-oriented software engineering IV. *Proceedings of the 4th International Workshop, AOSE 2003, Computer Science, 2935.*

Odell, J., Parunak, H. V. D., & Bauer, B. (2001). Representing agent interaction protocols in UML. In: *Agent-oriented software engineering. Proceedings of the First International Workshop AOSE-2000* (editors Ciancarini, P. & Wooldridge, M.). *Computer Science, 1957*, 121-140.

Sandholm, T. (1999). Distributed rational decision-making. In G. Weiss (Ed.), *Multiagent systems* (pp. 201-258). Cambridge, MA: MIT Press.

Searle, J. R., (1969). *Speech acts*. Cambridge University Press.

UDDI (2004), Universal Description, Discovery and Integration of Business for the Web. Retrieved from http://www.uddi.org/

Web Services (2004). Retrieved from http://www.webservices.org/

Willmott, S. N., Dale, J., Burg, B., Charlton, P., & O'Brien, P. (2001). Agentcities: A worldwide open agent network [Electronic version]. *Agentlink News, 8,* 13-15.

Wong, H. C., & Sycara, K., (1999). Adding security and trust to multi-agent systems. *Proceedings of Autonomous Agents '99* (pp. 149-161).

Wooldridge, M. (2002). *An introduction to multiagent systems*. Chichester, UK: John Wiley & Sons

Wooldridge, M., & Jennings, N. R. (1995). Intelligent agents: Theory and practice. *The Knowledge Engineering Review, 10*(2), 115-152.

Wooldridge, M., Jennings, N. R., & Kinny, D. (2000). *The Gaia methodology for agent-oriented analysis and design: Autonomous agents and multi-agent systems, 3*(3), 285-312. Boston: Kluwer Academic Publishers.

Endnotes

1 SOAP: Simple Object Access Protocol—SOAP is an XML (Extensible Markup Language)-based, lightweight protocol for exchange of information in a decentralized, distributed environment.

2 WSDL: Web Services Description Language—WSDL is an XML format for describing network services as a set of endpoints operating on messages containing either document-oriented or procedure-oriented information.

3 UDDI: Universal Description, Discovery, and Integration—The UDDI protocol creates a standard interoperable platform that enables companies and applications to quickly, easily, and dynamically find and use Web services over the Internet.

Chapter III

The HUB as an Enabling IT Strategy to Achieve Smart Organizations

Arturo Molina, ITESM, Campus Monterrey, Mexico

Ricardo Mejía, ITESM, Campus Monterrey, Mexico

Nathalíe Galeano, ITESM, Campus Monterrey, Mexico

Teresa Nájera, ITESM, Campus Monterrey, Mexico

Marcela Velandia, DaimlerChrysler Vehículos Comerciales México S.A de C.V., Mexico

Abstract

This chapter introduces the concept of virtual enterprise broker (VEB) supported by the use of a "HUB" of integrated e-services as an enabling IT strategy to design and create smart organizations. The VEB model is described in terms of core processes, success measures, and supporting ICT (information and communication technologies). The VEB is a business entity that enables the design, configuration, creation, and operation of smart organizations. VEB core processes are supported by e-services integrated in a "HUB" (The concept of HUB refers to a proposed centre of integrated e-services for virtual business) that is supported by Web-based applications and technologies. Six integrated e-services have been defined, based on the concept of on-demand services for value added industrial networks: e-marketing, e-brokerage, e-planning, e-engineering, e-supply, and e-productivity. The conjunction of these e-services improves industrial networks performance. A description of the e-services and HUB architecture is presented in detail.

Introduction

The concept of virtual enterprises (VE) has been around for some time (Byrne et al., 1993; Davidow and Malone, 1992; Goldman Nagel, & Preiss, 1995; Camarinha-Matos, Afsarmanesh, & Garita, 1998; Eversheim, Schuth, Bremer, & Molina, 1998). There are some frameworks that help to define how the virtual enterprise works. Table 1 shows the main characteristics of the models and research described in the literature related to virtual enterprises.

The research presented in this chapter has its foundation on the Framework for Global Virtual Business, created by COSME network (COSME was a network created by the European ALFA Project, between four European and two Latin American universities), to conceive how virtual enterprises could be designed, created, and operated to exploit new opportunities in global markets (Molina et al., 1998). This framework is described to some extent in order to explain the relevance of brokers in the formation of Virtual Enterprises.

The COSME framework defines three business entities: virtual industry clusters, virtual enterprise brokers, and virtual enterprises (Molina, Ponguta, Bremer, & Eversheim, 1998; Bremer, Eversheim, Walz, & Molina Gutiérrez, 1999):

- Virtual Industry Cluster (VIC) is an aggregation of companies from diverse industries, with well-defined and focused competences, with the purpose of gaining access to new markets and business opportunities by leveraging their resources and therefore their competences. The companies can be geographically distributed or not.

Table 1. Characteristics of the Models/Research in Virtual Enterprises

	Fundamental Concepts	Entities	Stages	Resources or Elements
"Dynamic Model of Virtual Organizations" [Lackenby & Seddighi, 2002]	VE: A temporary alliance between a number of core competence-based firms/individuals forced to take advantage of market opportunities	Cluster	1. Formation and Development Cluster 2. Leaving the Cluster 3. Assessing the environment 4. Determination of Structure Required 5. Operation Cluster	Core Competences
Virtual Corporations [Byrne et al., 1993]	The Virtual Corporation is a temporary network of independent companies, suppliers, customers, even rivals, linked by information technology to share skills, costs and access to one another's markets.	Entities roles necessary within the Virtual Corporation: Partnership, technology, trust, excellence (core competences), opportunism, and no borders.	The stages necessary to form the Virtual Corporation are not described. However, it is clear that the partners must complement each other's capacities and they must form a win-win relationship in order to obtain a specific market opportunity.	Core Competences Information Technologies
The Virtual Enterprise: concept of creating value [Katzy & Schuh, 1999]	The VE is based on the ability to create temporary cooperations and to realize the value of a short business opportunity that the partners cannot capture on their own. The VE is nimble: quick to grasp the new opportunity and create the solution.	Within this model, the roles rather than the entities are defined: Broker, Outsourcing manager, Network coach, Auditor, Project manager, and Competence manager	1. Create 2. Restructuring 3. Destroy	The Value that is the opportunity or reason restructuring takes place. The virtual operation or result of restricting. The network or source of restructuring
Framework for Global Virtual Business [Molina et al., 1998; Bremer et al., 1999].	The framework is defined in terms of three business entities: Virtual Industry Cluster, Virtual Industry Broker, and Virtual Enterprise. These entities are described in terms of its core products, processes, and core competencies.	Virtual Industry Cluster Virtual Industry Broker Virtual Enterprise	1. Identification of new business opportunity 2. Formation of industry clusters 3. Formation of virtual enterprise 4. Operation/Dissolution	Core competences Products/Processes Technologies
Virtual Enterprise Characterization [Goldman et al., 1995]	VO: is a dynamic organizational tool for agile competitors. It is at once neither temporary nor permanent. The VO must be customer-focused and opportunity-based, it must establish a set of world class core competencies to meet each opportunity.	The VO is composed by agile competitors, Goldman agrees with Byrne in the need of having opportunism, excellence, technology, no borders and trust.	Within this model the stages necessary to form the VO are not described. However, it is highlighted that the VO is an opportunity-based dynamic organizational structure. When the opportunity is over, the VO should be disbanded and celebrated.	Core competences Trust Open and honest communications Compatible management styles.
The Advanced Virtual Enterprise [Goranson, 2003]	Advanced Virtual Enterprise (AVE): best configuration of smaller players quickly aggregated to address an opportunity. AVE: Fluid supply chain facilitated by a lead partner	Partners Lead partner	1. High dynamic configuration 2. Change of partners and roles 3. Opportunistic formation/Dissolve/ Transition	Feature-based value metric Federation Mechanisms Auditable agent-based simulation and control capabilities Mechanisms to accommodate implementation realities, multilevel control, soft modelling and complexity management Standard methods to interface existing systems
Virtual Breading Environments [Camarinha-Matos & Afsarmanesh, 2003]	The VO main characteristics: Agility Complementary Roles Achieving Dimension Competitiveness Resource Optimization Innovation	Virtual Breeding Environments Brokers Virtual Enterprise	1. VE formation 2. VE Operation 3. VE Reconfiguration 4. VE dissolution	Agility Trust Building Breeding Environments

- Virtual Enterprise Broker (VEB) is a business entity that is responsible for searching opportunities in the global environment and enables the creation of virtual enterprises. The virtual enterprise broker performs the processes of partner search and selection, and configures suitable infrastructures for VE formation/commitment, that is, physical, legal, social/cultural, information. To achieve its goal the VEB supports its operations using a set of integrated information technologies.

- Virtual Enterprises (VE) are temporary networks of independent companies linked by information technology that share competences, infrastructure, and business processes, with the purpose to fulfill a specific market requirement.

These business entities work collaboratively to form a VE whenever the market needs it. Figure 1 shows the interrelation between those entities. The VEB is presented as the business developer, the VIC is shown as a congregation of enterprises of defined sectors, each sector complementary to one another, and finally the VE is outlined as the dynamic entity that enables the formation and disintegration in accordance with market requirements.

Industry clusters are regional networks of industries with common product chains. Based on this concept, VICs are created by the aggregation of enterprises around the world with complementary competences, supported by information technology infrastructure. The use of information technology overcomes the restriction of geographic proximity of industry clusters. Therefore, virtual industry clusters based on a global network of best-in-class companies can be created.

Figure 1. Framework for global virtual business (Molina, Molina, Ponguta, Bremer, & Eversheim, 1998)

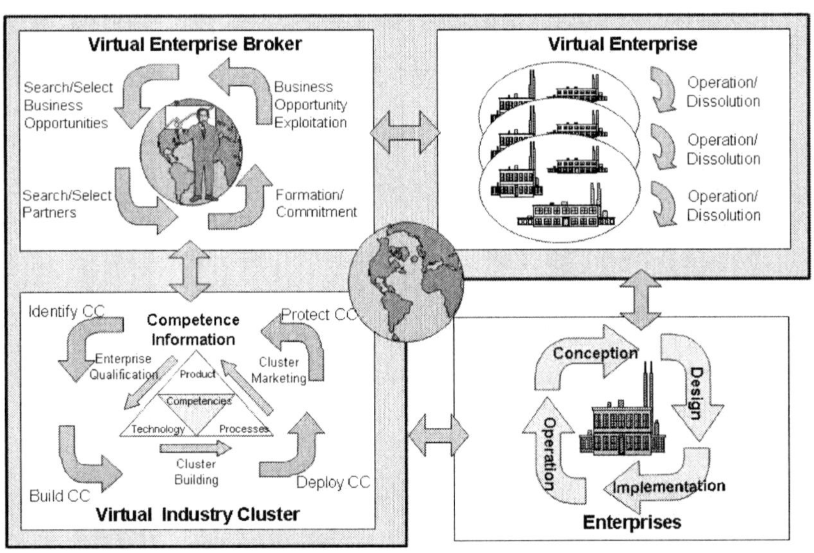

The core products of VICs are the collection of all the human and technological resources and processes that can be offered by the cluster. The commercial success of VIC depends on how well-defined and focused the cluster is (Molina, Flores, & Caballero, 1999). Hence, the information regarding the technological resources, processes, and human capital has to be structured in a manner that can be used by VEB to search and select partners for VE. Furthermore, this information should be used to support strategic decision-making in managing the VIC's core competences.

The VIC has three main core processes: cluster definition and design, cluster creation, and cluster core competences management. The first two processes are related to the organization, formation, and marketing of the VIC, which includes searching, selecting, and qualifying enterprises (Caballero, Molina, & Bauernhansl, 2000). The latter involves identification, building, deployment, and protection of the VIC's core competences (Galeano, 2002). The VIC success relies on the effective management of its own core competences and the marketing of the competence aggregation of the members. The VIC's core competences are the aggregation of the competences of its members, the ability and flexibility to integrate and deploy resources and capabilities, and an effective cost management. This aggregation should have a focus, and therefore should represent the competence of the VIC. The competences of a VIC can be the capacity of the cluster to make certain types of products, offer a group of business processes or provide specific technological capabilities. The results of research carried out in the design, creation, and implementation of virtual industry clusters can be found in Flores and Molina (2000).

The virtual enterprise broker (VEB) exploits business opportunities through the creation of virtual enterprises. Ávila, Putnik, and Cunha (2002) have defined broker's functions as explicit or implicit. The explicit functions are related to the initiation of the virtual enterprise, resource selection, integration, and configuration while the implicit are related to the interaction with another brokers, resource maintenance, negotiation, and transactions. All these functions are defined as core processes for the VEB, and therefore must be supported by information and communication technologies (ICT). The core processes of a VEB are related to the organization and deployment of competences of VIC's members in order to create, organize, and integrate partners in a VE. Therefore, key business processes are opportunity search, partner search, and VE formation and commitment. The VEB has also the responsibility of configuring the adequate infrastructure for the successful operation and dissolution of VEs. The most important core competence that a VEB should posses is the ability to integrate the competences of partners into successful VEs that meet customer's requirements. In order to be successful in the global business environment, the VEB has to build its own competences concerning competitive advantages, strategic focus, and technological capabilities. The VEB should decide how a VE will differentiate from its competitors; for example, by providing complete product chain processes, taking geographical advantage from suppliers, procuring best business practices, or specialized technology (knowledge engineering, high-end technology).

The formation process of the Virtual Enterprise involves the creation of partnerships between companies in VICs. However, in each activity the VEB has to be a participant in such a way that it serves as initiator, driving force for the formation of the VE, and coordinator through defining the business arrangements and assignments of the VE memberships, and serve as moderator during the executions/operation of the VE's mission by resolving conflicts between VE members (Kanet, Faisst, & Mertens, 1999). The VEB needs to have access to sufficient and explicit information about several domains in order to be able to form a particular VE (Harbilas, Dragios, & Karetsos, 2002), for example, it requires competencies to integrate and to deploy enterprise capabilities and capacities, to manage projects and to develop products, processes, and manufacturing systems. In the proportion that the VEB manages the technical or technological knowledge to run a project, the core competences and processes of the partners will be complemented. In the next section, a detailed description of the Virtual Enterprise Broker is included in order to define how smart organizations can be created using this model.

A Virtual Enterprise Broker Model to Create Smart Organizations

Definition of a Smart Organization using the Global Virtual Business Framework

In our context of industrial networks, the cooperation model of company alliances is moving from *virtual enterprises/virtual organizations*, to what is nowadays called *smart organizations* (Bultje & van Wijkt, 1998; Filos & Banahan, 2000; Ferreira, 2001). Specifically, RoundRose (2004) describes smart organizations as:

- Internetworked
- Virtual in concept
- Dynamically adaptive to new organizational forms and practices
- Knowledge and learning-driven
- Hierarchically flattened where the individual's skills, intellect, and knowledge are recognised, valued, and leveraged

Smart organizations involve more than the capability of setting up a digital infrastructure and the ability to enter into virtual collaboration with other partner organizations. The approach presented in this chapter aims to achieve the creation of smart organizations through the use of the following elements: virtual industry clusters, virtual enterprise broker, and an ICT infrastructure called "HUB". The HUB concept is a center of integrated e-services for Virtual Business to create value added networks of small and medium enterprises. In the framework for global virtual business, the virtual enterprise broker enables the creation of new business through the aggregation of core processes, competences, and information technologies of enterprise networks (virtual industry clusters). Figure 2 shows how the combination of virtual industry clusters and virtual enterprise brokers can conceive smart organizations. The core competence of a smart organization is the aggregation of business

Figure 2. A smart organization based on the framework for global virtual business

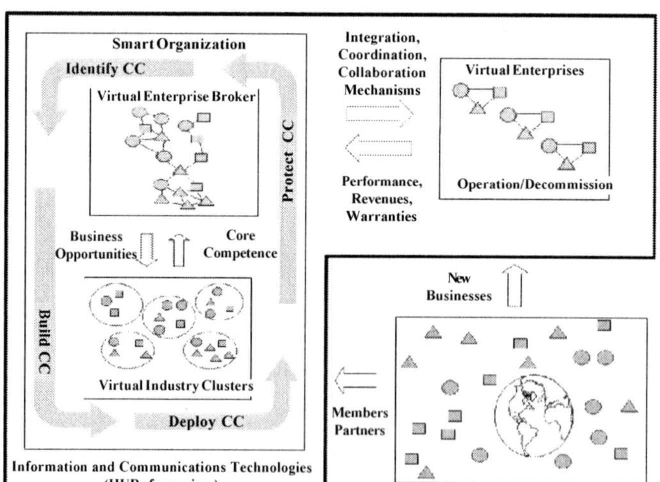

competences of the VEB and technological competences of the VIC that should be able to create added value to customers. This aggregation of business and technological competences, together with the information technologies supporting the operations of the VEB, enables the creation of a smart organization. This conceptualization of a smart organization is applicable to different kinds of companies, industries, or networks.

In a smart organization, major emphasis should be placed in managing core processes and competences. The management of the smart organization now should combine the management of the competences of the VEB, as well as the competences of the VIC, to build and sustain the core competences of the smart organization. The following section describes a core competence management model for smart organizations.

A key core competence in the VEB consists of information technologies that must be designed, integrated, and managed by the VEB in order to ensure that each core process will be supported by the right technology. An integrated HUB of e-services plays an important role in the brokerage services offering support for the operations and services required for the organization, creation, and operation of smart organizations.

A Core Competence Management Model for the Smart Organization

The model for core competence management for smart organizations includes four main strategic processes:

- Core Competence Selection
- Core Competence Development
- Core Competence Protection
- Core Competence Deployment

In the core competence selection the companies with the required core competencies for the smart organization are identified by the virtual enterprise broker. Different virtual industry clusters can be created based on the companies' technological competencies. Once the enterprises are identified and selected, the activities of the cluster formation are performed; enterprises are evaluated and qualified, and information databases related to the core competencies of the different clusters are organized.

The core competencies of enterprises belonging to the smart organization can be improved, enhanced, or new core competencies can be developed. The set of activities for core competence development are evaluation of company performance, best practice performance analysis, design of enterprise development plan, and execution of process improvements. The outputs of the core competence development process are projects developed in the enterprises of the smart organizations and the identification of complementary core competencies to strengthen the current technological capabilities of the virtual industry clusters.

The main objective of the core competence protection is to achieve the sustainable competitive advantage of the smart organization. It includes the evaluation of different aspects during the VEB and VIC operation: analysis of the clusters performance (operating efficiency, financial self-sufficiency, coverage, and effectiveness), analysis of enterprises performance (enterprises member participation, commitment, and response during project execution), financial and legal aspects (which should be analyzed in order to work within law regulations, contracts, warranties, intellectual property rights), and finally, the requirement for information and communication technologies to improve the organization operations. This last issue allows the continuing development of electronic services integrated in a HUB that are shared among the virtual enterprise clusters, virtual enterprise broker, and virtual enterprises.

The core competence deployment focuses on the use of the smart organization competencies to achieve new business opportunities. During the deployment, market demands and business opportunities are identified, all the business opportunities are analyzed by the VEB, and if the enterprise core competencies in the VICs match with the customer requirements, a negotiation can begin and probably a contract can be made. The deployment includes product or service delivery to the customer and the after-sales customer service.

The activities of the core competence model—selection, development, protection, and deployment—is a cycle that defines the VEB competence management, and therefore a key competence in the smart organization. The core competence deployment is a set of operational core processes that the smart organization should carry out to design, organize, create, and operate Virtual Enterprises. In the following section, these core processes are introduced.

Operational Core Processes, Measures, and Information Technologies

The smart organization operational core processes are the ones that the VEB has to perform to be able to satisfy a customer requirement. The VEB core processes are described in Table 2 (Molina, Mejia, & Velandia, 2003). In this chapter, the focus is on the information technology (e-services) required to support the VEB operations, in order to ensure such added value in the delivering of products and services. The ICT must support each of the core processes and must be shared among the VIC, VEB, and VEs. These information technologies are described in six e-services: e-marketing, e-brokerage, e-planning, e-engineering, e-supply, and e-productivity.

The VEB must be measured as any other entity in the Virtual Enterprise. However, the measures used to assess the VEB are more critical than those in the partners. This is because the VEB manages the entire supply chain and little waste of resources can be multiplied over the chain. If a partner does not demonstrate reasonably good productivity at the end of a project, it could

Table 2. VEB core processes, measures, and e-services

Core processes	Description of Core Processes	Measures of Processes	e-services
Search and Select Business Opportunities	Search for business opportunities Business Intelligence analysis Customers requirements detection	Number of new liaisons Number of new business opportunities	e-marketing
Negotiation	Reception of project requirements Analysis of required technological competencies Search and selection of partners Request for Quotation Commitment and Binding	Number of RFQ delivered on time Number of RFQ approved /Total of RFQ Value of Contracts/ Total of RFQ	e-brokerage
Project Planning	Project plan development Confidentiality agreements Negotiation of deliverable times, monitoring and inspections Contracting	Time for preparation of project plan Amount of resources involved / Amount of resources planned	e-planning
Project execution	Project coordination: management and control Project execution: collaboration, supervision and inspection. Product realization, order fulfillment process or customer service	Productivity: value added per supplier Efficiency: cost per supplier Effectiveness: deliveries on time	e-engineering e-supply
Customer Follow-up	Customer audits and feedback Performance analysis of enterprise network Profit sharing Warranties responsibility Contract expiration	Number of customers that repeat an order/ Total number of orders Number of claims per customer	e-productivity

impact in the delivery time or planning of the project. However, if the VEB does not show the productivity expected the results can be a negative ROI. This is the main reason why information technologies help the VEB comply with the objectives throughout the VE life cycle.

In project planning, the performance is compared to the effectiveness of planning in accordance to execution. Moreover, in project execution, the measures are used to determine whether time, cost, and conditions requested by the customer are achieved. These requirements are generally improved by the use of tools like e-engineering that will improve communications between the VEB and the final customer, from the design to the manufacturing phase. E-supply is also a helpful tool during project execution, due to the high integration required in the supply chain.

Finally, in customer follow-up phase, customer relationship managmenet (CRM) technologies are helpful to follow customer requirements and identify new business opportunities. Moreover, e-productivity helps to measure the performance of partners involved in the project, in such a way that this information will be key for partner selection in future projects.

HUB of Integrated E-Services to Support Broker Operations

New information technology is necessary for small and medium enterprises (SMEs) where the creation of value added industrial networks allow companies to share technological competencies in order to have access to new global business opportunities. These networks require integrated information services (e-services) that offer and enable coordination and cooperation among the different SMEs. These services allow companies to share their knowledge and technological resources to create virtual and smart organizations. These e-services should be integrated in an open technological platform, easy to access, known as "HUB." The HUB described in this chapter is conceived to offer SMEs services in such a way that foster the creation of smart and virtual organizations.

But, how the VEB can achieve smartness in a virtual organization? The combination of different parameters, as defined in a previous section of this chapter, makes the broker an enabler of virtual businesses through the exploitation of global business opportunities. These parameters taken from the smart organization definition (Ferreira, 2001) are the starting point to achieve smart organizations, where the business environment is composed of value added networks (clusters of internetworked organizations) collaborating around a particular technology and making use of a common architecture (HUB) to deliver independent elements of value (e-services) that grow with the number of participating organizations (SMEs and original equipment manufacturers).

Then, the smartness of the organization is implemented to support market changes, taking advantage of the ability of configurability and flexibility. The variability of products in today's markets is increasing due to mass customization and one-of-a-kind production, forcing the smart organization to be prepared and have the ability to rapidly configure its competencies in order to fulfil customers requirements. This attribute can be achieved through the configuration and integration of capabilities and capacities of different companies, which is the main issue of the VEB operation model. However, the ability to foster this internetworking environment among companies should be supported by a HUB of integrated e-services, in order to offer a common architecture to the companies and their clients through a set of services to be used on demand.

The concept of creating a HUB (Integrated e-services center for virtual business) is considered a "technological innovation" because it enables the integration of value added networks, creating virtual and smart organizations. This concept reduces critical troublesome that limits SMEs competitiveness, allowing new business opportunities to be exploited and having access to new global markets. The HUB offers a variety of integrated e-services that are necessary for dynamics market competition. Innovation presents three variants:

- HUB development with Internet-based services, allowing SMEs to have access to a wide range of value added e-services.
- e-services implementation methodology, to demonstrate its impact and benefit for the SMEs through an integrated process to achieve competitiveness.
- Creation and demonstration of the new SME business model based on value added industrial networks that enable the creation of smart organizations.

The integrated HUB should not affect the normal flow of other services, but it can increase their commercial potential, being able to integrate them to the e-services portfolio for SMEs. Some commercial applications are considered to integrate into the HUB that might be configured for SMEs, such as: CRM (customer relationship management systems), EPS (electronic procurement systems), SCI (supply chain integration), LMS (logistic management systems), and PLM (product lifecycle management), among others. The key issue is to offer to SMEs the possibility to use the e-services, which support their business processes at an affordable cost on demand services.

Integrated E-Services in the HUB

There are five integrated e-services (as shown in Figure 3) that a HUB should offer as core for on-demand e-services for value added industrial networks: e-marketing, e-brokerage, e-engineering, e-supply, and e-productivity. The conjunction of these e-services should improve the competitive position of SMEs based on the virtual organization concept.

Figure 3. HUB of integrated e-services supporting the smart organization operations

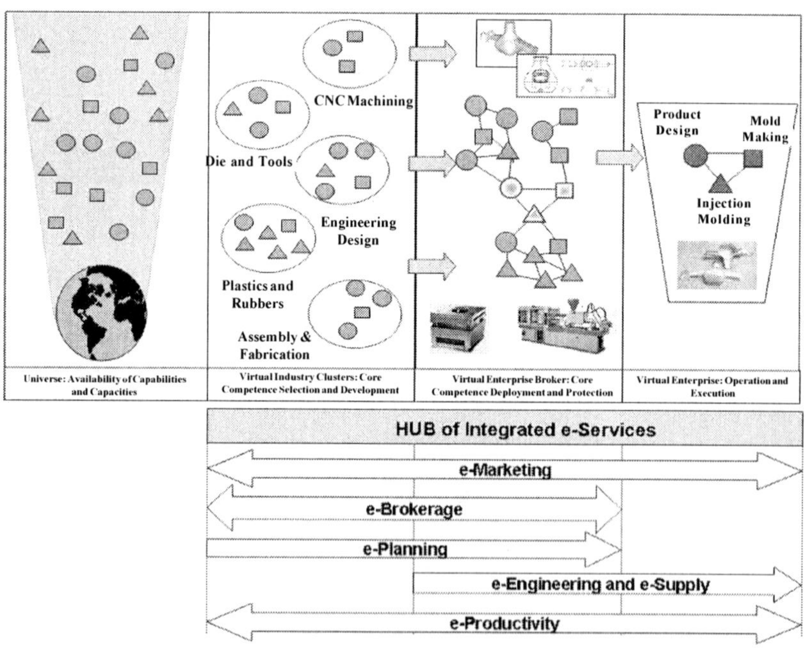

The HUB is center of integrated e-services to support the design, formation, and operation of smart organizations of SMEs. The need for integrated e-services emerges from the limitation of individual current electronic services, which do not offer a solution to the current technological limitations of SMEs. The e-services and their conceptualization are presented in the next sections.

E-Marketing

The combination of the 4P's of marketing—Product, Price, Place, and Promotion (Ketle, 1997)—have lead to the creation of a new e-marketing concept where product price, sales channel, place of the producer, and promotion tactics have become the critical content of market analysis and management tactics in enterprises. According to Qin, Xi, and Chen (2001), the innovation of e-marketing is the tactics combination from four aspects:

- **The Innovation of Product Tactics:** Customer service must be done online, customers should have constant service, and customer behavior must be fedback immediately to the companies.
- **The Innovation of Price Tactics:** Customers should be supported to compare products and processes over the network. The customer should always get the best deal based on the customer critical factors: quality, price, time, volume, and flexibility.
- **The Innovation of Marketing Channel:** Customer should gain access to all possible markets (vertical and deep) using one channel.
- **The Innovation of Promotion Tactics:** Customer must overcome the limitation of time, space, and region, e-marketing should cover the whole world, conveying multimedia information.

Therefore, based on the mentioned tactics, e-marketing should offer a renewed way of promoting virtual organizations through the Internet by offering a set of services based on the following possibilities (Evans & King, 1999):

- Multiple marketing uses,
- Access to commercial research and competitive intelligence activities,
- 24/7 customer service,

- The Web as a sales channel,
- Support for channel partners,
- Around-the-clock presence,
- Up-to-the-minute information,
- Linkage with partners, and
- Interactive multimedia vehicle.

Usually, SMEs have inadequate products and services commercialization in international markets. The limited access to competitive intelligence information for competing in global markets is also a restriction that SMEs have to overcome in order to exploit new market opportunities. E-marketing service offers an intelligent portal that will allow the promotion of products and services facilitating management and maintenance activities for the SMEs.

The e-marketing concept is oriented to the ownership of SMEs of their own Web portals, in such a way that content updating can be done in a simple way by the SME. Wizard-based prototypes have been developed to enable fast configuration and updating of Web portals through the concept of reconfigurable and modular software. The concept of smart portals is included also, based on competitive intelligence analysis tools of current and potential customers. It integrates technologies for electronic catalogues customization and customer service management.

E-Brokerage

Brokerage service is a mediation process between customers and suppliers. It is formed by various types of processes, supporting the matchmaking between consumer needs against suppliers' capabilities and skills. It is a communication channel for negotiation and bidding; the following processes are supported:

- **Information Management Process:** To structure and dynamically aggregate information of supplier and users registered in electronic catalogues; that is, to enable users the appropriate view on the supplier/customer domain.
- **Offer Communication Process:** Enabling a supplier to provide information regarding its capacities, capabilities, and resources to offer products

and services, and through which a consumer can communicate an interest in products and/or services.

- **Negotiation Process:** Supporting negotiations (e.g., bilateral and multilateral) on products and services. Negotiation support is considered one of the most difficult forms of matchmaking mechanisms and is a core focus of this e-service.
- **Quotation Preparation Process:** Offering bid preparation capabilities based on user requirements matching suppliers' capabilities taking into consideration key aspects such as: quality, volume, time and price.

Important tasks to be automated in e-brokerage services include the following (Bichler & Segev 1999):

- Product/services taxonomies, ratings, and statistics.
- Product comparison and selection.
- Notary services like nonrepudiation of offers, orders, and contracts.
- Support for billing, accounting, and digital delivery.
- Advanced payment services like a letter of credit.
- Certification of liability and reliability of participants.

Therefore, the e-brokerage service is based on the brokerage process required to create value added industry networks. Brokerage services are required to be modelled and simulated in order to demonstrate business scenarios before they are implemented, allowing the design of the best possible configuration of industry supply nets. The dynamic of industry supply nets must be supported by a comprehensive collaboration environment, which allows different companies to interact during the complex process workflows of negotiations. Some specific services that could be offered by the HUB that support brokerage operation are (Mejia, Aca, Garcia, & Molina, 2002):

- **e-selection:** This tool facilitates the identification, evaluation, and selection of a partner according to certain capabilities or resources needed to exploit a specific business opportunity. The selection uses past performance data of companies' behaviour in industry networks.

- **e-RFQ (Request for Quotation):** The main purpose is to reduce the time consumed during the quotation process, and to facilitate the communication between customer and suppliers networks
- **e-bidding:** This application allows preparing a collaborative bid based on a value added network of companies.

E-Planning

Planning, design, and operation (management) goals and requirements of smart organizations are generally different from those of single, centralized enterprises. The basic feature of a smart (virtual) enterprise is that its cooperating units keep their independence during the life cycle of the cooperation (Kovács & Paganelli, 2003). The e-planning service intends to help customers and providers to define the work plan prior to the execution of a project. Tools such as scheduling, cost estimating, and resource balancing, among others, should be provided in this service. However, companies are unwilling to install a lot of applications on their computers, increasing the potential to centralization through a common database for project planning. A solution to this problem can be achieved with Web-based applications. Companies are interested in providing their customers with light, custom Web-based applications. For this reason, the management practices that normally are executed in a project planning activity are described to be analyzed in order to define potential processes to be automated in an e-planning environment:

- **Contracting:** The actual formation of a project, the purposeful, formal initiation, negotiation, and agreement of a project.
- **Controlling:** The process of monitoring and steering the performance of a project, including accounting, budget control, evaluation, and allocation of resources.
- **Stratifying:** The process of cultivating the pools that make up project networks; decisions on who is inside or outside a project network and/or the core of a pool.
- **Socializing:** The process of attending networking events to cultivate contacts, create ideas and identities, share information, and manage expectations.

As mentioned by Panta, Sethia, and Bhandari (2003), a key element of the cooperative planning and forecasting mechanism is a meta-supply chain model residing on a common Web application server, provided by a neutral party and configurable by the contracting parties involved. However, the specific models (workflows and best practices) should be configured by each company, because the tactical planning model for any party consists of information that is specific to its situation. If the situation with any of the parties changes, since they are all interconnected, the revised information could be communicated in real time to the systems of other parties to enable them to modify their respective plans. In an integrated e-supply chain system, disruptions should be incorporated in the customer's production schedules instantaneously.

E-planning can be linked to other e-services, such as e-engineering. Every e-engineering process requires a project planning and a project execution process. The planning process can be achieved the same way as working through an e-planning environment; this means gathering all the information previous to project execution (including real-time negotiation, risk management plan, and changes during project management plan), monitoring the process, and assigning tasks to the project participants. This e-planning concept has been defined in collaboration with the European project e-HUBS (e-engineering enabled by Holonomic and Universal Broker Services, IST-2001-34031), which has developed a technological platform for engineering project planning and negotiation.

In consequence, the e-planning can offer services for contracting using real time negotiation, legal aspects, project schedule, and deadlines establishment. This can be achieved using systems such as e-Legal (Hassan, Carter, Hannus, & Hyvarinen, 2001), which consist of a negotiation process using a "virtual negotiation room" on the Internet in which different parties are guided to configure a contract automatically. Some other services like project or task execution supervision and monitoring can be offered, assuming that one result of an e-planning service can be the activities breakdown. This service can be complemented with business process management definitions and its monitoring techniques, in order to control project progress and track potential problems.

E-Engineering

Small- and medium-sized companies play one of the most important roles in the economy. However, especially in Latin America, they have the following characteristics: (1) highly experienced engineers who lack knowledge about new technologies, and (2) a lack of infrastructure (hardware) and engineering tools (software) to offer highly skilled engineering services. The main issue to tackle with this e-service is the limited capability of companies (especially small and medium enterprises) to design, develop, and innovate products. Those SMEs lack properly implemented product development processes and qualified engineers in design and manufacturing engineering.

Collaborative e-engineering environments emerge as a key issue to provide resources, services, and knowledge at low cost for engineering tasks. The e-engineering service is based in the design and integration of those environments for SMEs to support their engineering activities. The HUB can offer the service of collaborative e-engineering environments configuration for the specific processes of each company (customization), allowing companies to focus on their core competencies, instead of taking care of technological issues. Under this concept SMEs will not require the technological capabilities (hardware) that specific engineering applications require. Engineers can focus on their engineering activities regardless of technological knowledge. The benefits for SMEs will be reflected in terms of savings due to lower costs than having their own infrastructure with high maintenance costs and specific technological knowledge.

Collaborative e-engineering environments for integrated product development rely on the implementation of Web technologies and applications (Aca et al., 2003) to foster collaboration among engineering partners and to support engineering activities for product design and manufacture. This environment would cover four main technological areas: *Functional tools* (CAD/CAM/CAE stand-alone tools and Web-based applications developed to support specific engineering activities); *Collaboration tools* (net-meeting, forums, chats, multicasting, e-mail, and shared spaces applications); *Coordination tools* (Process models, workflows, and project management software); and *information/knowledge management tools* (product and manufacturing models, PDM, and file repositories). Due to this, the e-engineering service provides the user companies an alternative to coordinate and collaborate between them and their clients and suppliers, through the use of a collaborative e-engineering environment.

Finally, the e-engineering service intends to offer the companies methodologies and processes to develop new products, through the use of workflows as a guiding tool for product development process management. The activities proposed will be based on defined frameworks, introduced as best practices in design activities. This model should integrate the relevant activities normally carried out in product design, in order to offer to SMEs a structured way of designing, innovating, and developing their products.

E-Supply

Currently, there is a huge amount of information in manufacturing companies that is shared between business processes among the supply chain. An explicit necessity is to manage this information under a structured approach that integrates this information in order to improve the efficiency of the key business processes and the performance of the entire supply chain, from the supplier to the customer (internal and external customer).

For these needs in supply chains, a solution to overcome these problems and to simplify business processes could be a Web portal. Interactions with one Web portal are easier to manage than many peer-to-peer relationships, as expressed in Boyson, Corsi, and Verbraeck (2003). A Web portal can provide unified database, linked across all functional systems, both within the organization and among the organization and its major supply chain partners.

One of the most common barriers to full supply chain management has been the cost of communication with, and coordination among, the many independent suppliers in each supply chain (Fredenhall & Hill, 2001). Information technology makes it possible to digest, to understand, and to act on this growing abundance of information by using sophisticated analysis, modeling, and decision support capabilities. Internet technologies allow companies to demand data and supply capacity data to all companies within a networked supply chain. Due to this, companies can forecast demand fluctuations and have the ability to rapidly respond to market requirements, due to updated capacity data, shared by all companies in the networked supply chain. Additionally, as expressed by Boyson Corsi, Dresner, and Harrington (1999), Internet-enabled shared information helps break down organizational policies and functional fences, helping supply chain alliance members develop a common understanding of the competitive environment.

Firms are increasingly embracing integrated Web-based or e-supply chains because such chains are believed to enhance efficiency and competitiveness. The e-supply chain movement has received a boost from a variety of off-the-shelf supply chain software solutions that have appeared on the market. However, in the excitement about these software solutions, it is often overlooked that creation and implementation of integrated supply chains require tremendous resources, a great deal of management time and energy, large, organization-wide changes, huge commitment from suppliers and partners, and sophisticated technical infrastructure (Panta, Sethia, & Bhandari, 2003).

The e-supply service integrates technologies for the optimization of logistic systems and supply chain integration, and enables import/export services for products and materials and supply chain management.

The principal guidelines that enable this service are the customer requirements, supplier capabilities, product characteristics, and available resources. E-supply chain management can be categorized into three areas: supply chain planning, supply chain execution, and supply chain transaction (Reddy & Reddy, 2001). The integration of applications to improve the sourcing process is the issue to be tackled. One first approach in this project has been the development of a manufacturing execution system, which integrates the three areas described above in a Web-based system based on open source technology (Ramirez-Santaella & Molina, 2004).

E-Productivity

E-productivity incorporates technologies for diagnosis, planning, and monitoring of SME development according to productivity and benchmarking indicators. An application has been developed for diagnosis, planning, and monitoring SME development (http://cax.mty.itesm.mx/immpac), including a benchmarking with similar companies (Molina, Gonzalez, Galeano, Flores, &Caballero, 1999).

This application is a measurement tool structured in four stages. The first one is related to the competitive position—the characteristics of products, markets, clients, suppliers, and processes of the companies are identified. The second stage measures productivity indicators in terms of quality, volume, time, profitability, flexibility, and innovation. This stage establishes a system of indicators for its continuous evaluation. The third stage analyzes the manufacturing practices implemented in the company and the level of implementation. The fourth and final stage analyzes the information gathered in the previous stages and formulates a plan of enterprise development. In this stage, simulation tools are used in order to validate the impact of the development plan. These simulation tools simulate the effects of changes in manufacturing operations, and help to prioritize potential improvement projects, allowing the company to explore the consequences of "what if" scenarios under a modelling environment.

The use of benchmarking as a core tool for e-productivity allows an organization to monitor environments for competitive practices (Camp, 1989; Ulrich, Brockbank, & Yeung, 1989; Walleck, O'Halloran, & Leader, 1991; Karch, 1992; Main, 1992; Jacobson & Hillark, 1986). Therefore, companies can use this tool to measure their productivity level and compare their improvements in business and manufacturing practices.

E-Services Implementation Architecture

Technological architecture plays an important role for e-services success. The middle step between concepts and reality is the architecture definition. With a model definition, implementation can be performed based on the technological requirements and the distribution of the different modules involved in the HUB. E-services are represented as business processes that can be externalized via Web services. This architecture evolves from traditional n-tier architecture to a service-oriented architecture, offering a standard-based value proposition.

Business process management systems (BPMS) provide not only the tools and infrastructure to define, simulate, and analyze business process models, but also the tools to automate business processes as workflows (Leymann, Roller, & Schmidt, 2001).

The BPMS infrastructure provides the run-time environment for the e-services. It allows users to monitor the execution of individual processes, to analyze the overall behavior of a set of business processes, to verify their successful performance, and to provide input for process optimization.

The use of BPMS provides the HUB with the capability to configure (assemble) any e-service for a specific client on the fly, without the need to write code for the new service into the application (taking into consideration that the e-service can be completed with the functionalities available to the HUB; otherwise, additional functionality must be added to the HUB). Once the e-services are assembled, the HUB can provide them to its clients through the use of Web services.

Web services provide a standard means of interoperation between different software applications, running on a variety of platforms and/or frameworks (W3C, 2004). The Web services set of standards and technologies represent the evolution of past distributed component technologies like remote procedure calls (RPC), ORPC (DCOM, CORBA, Java RMI), messaging (MSMQ, MQSeries), and even modern Web applications (Skonnard, 2002). Like their predecessors, Web services are intended to enable application programs to communicate and share functionality across networks. However, Web services are designed to leverage the standards-based architecture of the Internet and World Wide Web to promote interoperation across a wide variety of computing platforms. The three primary Web services standards (Roy & Ramanujan, 2001) are Web services description language (WSDL), simple object access protocol (SOAP), and universal description discovery and integration (UDDI).

Web services architecture is based upon the interaction among three roles: service provider, service registry, and service requestor (IBM, 2001). The interactions involve operation publish, find, and bind. A services provider hosts a network-accessible Web service (a software module with specific functionality), which is an implementation of a Web service. The service provider defines a service description for the Web services and publishes it to a service registry. The service provider can also publish it to other places, including directly to a service requestor, if desired. The service requestor uses a find operation to retrieve the service description locally or from the service registry, and uses the service description to bind with the service provider. The requestor then binds with the service provider, and invokes or interacts with the Web service.

The HUB architecture model shown in Figure 4 is an example framework covering the necessary foundation for building and deploying e-services. This includes four layers: Enterprise infrastructure, Web service infrastructure, Web infrastructure, and Internet.

Enterprise Architecture

Is the collection of strategic and architectural disciplines that encompass the information, business system, and technical architectures (Perks and Beveridge, 2003). In other words, it's the layer that holds the business models, operating systems, databases, applications, and the services.

Service Infrastructure

Service infrastructure is composed of a service broker, a Web service BUS and BPM.

Figure 4. HUB architecture model (Adapted from Trivedi, 2002)

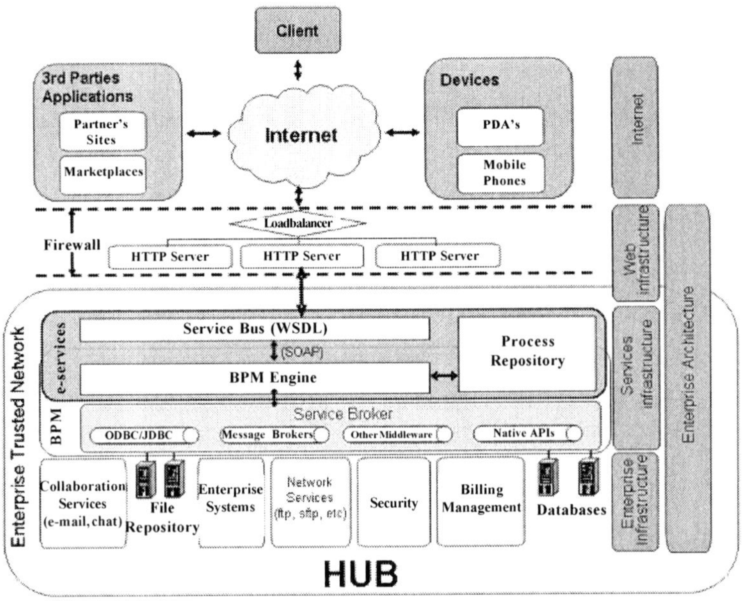

The service broker handles e-services requirements for information and processes execution from one or more applications. The communication with and between all the applications is done with one or more of these tools: ODBC, APIs, Middle Ware, Message Broker, and Web services.

The Web service BUS (WSDL) from where internal and external applications consume the e-services.

The BPM, which encloses the definitions of the business processes (by means of a process repository) for each e-service, executes the steps of the process in response to a request from a Web service or as stated by the business process. If a step requires human interaction it will wait for a specified time for that step; if the time is exceeded, it will issue the necessary signals (e-mail, SMS, etc.) to get the attention to complete the step. During the execution of the business process, the BPM must take into consideration other aspects such as billing, the security and legal issues regarding the use of the information, and its impact in transactions.

Billing is the mechanism by which the HUB will be able to charge for the usage of the e-services. How the charges are made is described in the business model for each e-service. Billing deals with credit card processing, transaction validation through online bank statements, and invoicing.

Security mechanisms must be implemented for the interchange of information by the services and the applications. This includes access to the OS layer (basic I/O, network services, and file systems), access to databases, access to applications, and access to devices. Given the legal value that is often attached to data managed and exchanged during the execution of an e-service, it is required that the HUB logs every transaction it executes.

Additionally, the whole area of security issues, from the more basic ones (availability, authentication, integrity, confidentiality) to the more complex ones (authorization, non-repudiation) is an equally critical aspect to be able to trace down responsibilities of "who did what."

Web Infrastructure

Web infrastructure can be seen as the gateway between the clients and the HUB. It provides the security to protect the HUB from the Internet (Firewall), takes care of Web encryption protocols like Secure socket layer (SSL) for the client connection to the HUB services, and the load balancing on client access to the HUB.

Internet

The Internet represents the clients that use the HUB e-services. This can happen through a Web page or by a direct access to the Web service. The client can be a person accessing the e-service through a Web page or an application front end; or another Web server that encapsulates the Web services with a particular look and feel for its clients; or an automatic process that interacts with an e-service.

To better understand how all this infrastructure comes together to offer e-services, a functional description of the e-services execution in the HUB will be detailed. The following set of steps will describe how to trigger the process, how the process is performed, and how the information returned to the requestor:

- **Step 1:** A request that can be through a predefined URL or a Web form is submitted to the Web server by the user.

- **Step 2:** The Web server then starts the execution of the corresponding application specified by the request.
- **Step 3:** The application obtains the WSDL from the Service BUS and executes the Web service. The Web service triggers the BPM engine to start the process.
- **Step 4:** The BPM engine, from the process repository, obtains the process definition. The tasks of the process can be synchronous (stops the processing until the step is done) or asynchronous (can go on to execute other steps without the current step finishing).
- **Step 5:** The Web service returns the processed information to the application.
- **Step 6:** The application transforms the information into a Web page and returns it back to the user, ending the request.

The first three steps are common to most of the e-services. Step 4 is also common in terms of BPMS process execution, but its specific tasks depend on the requested service. This is the basis of the BPM engine flexibility to achieve integration to different types of programs, allowing it to accomplish nearly any task. Steps 5 and 6 are common to most of the e-services, returning the result of the e-service and the confirmation of its execution.

The combination of BPM and Web services allows this HUB to easily describe services and orchestrate the interaction of their associated tasks. The services can be consumed by external applications or users obtaining benefits, such as easy integration and low cost through the use of outsourced information and communication technologies.

Conclusion

The creation of smart organizations achieved through value added networks using a brokerage role seems to be a competitive model for SMEs. However, it requires more advanced technological developments to enable the integration of e-services in the form of a HUB of integrated core business processes. The HUB offers a solution to SMEs's technological limitations in a systematic and integral way, through accessible costs of IT capabilities. Several pilot applications have been developed to become e-services, however the integrated architecture and infrastructure of the HUB is under development.

The competitive advantage of the HUB is based on:

- E-services focusing on the SMEs core processes, required to increase their competitiveness through commercialization of their products, competitive intelligence analysis, and negotiation, support their product development process and improve their input/output logistics and business management.
- The creation of an open architecture based on BPM and Web services, enabling the integration of proprietary applications, as well as "off-the-shelf" software and services already offered by other companies (e.g., Web pages publishing, Directories, Vertical Markets, etc.)
- The concept of "unique e-services hatch" through the integration of a set of e-services in a unique service provider (HUB) who allows offering multiple services according to the identified needs in SMEs.
- The e-services set up with a low investment cost for the SMEs, for exploring and creating new business based on the concept of virtual and smart organizations.
- An "on-demand" schema offer of services that are continuously requested by SMEs.

Acknowledgments

The authors acknowledge the chair in mechatronics from the Instituto Tecnológico y de Estudios Superiores de Monterrey (ITESM - Campus Monterrey), and the "IBM SUR Program" from IBM, for their support in the development of this research.

Authors also acknowledge the Inter-American Bank of Development for funding the PyME CREATIVA project (ATN/ME-7956-RG-11), and the ECOLEAD project (FP6 IP 506958). A special thanks to its research team for their contribution in the conceptualization of the different e-services defined in this chapter.

The e-planning pilot project is part of the e-HUBs project (IST-2001-34031) sponsored by the European Commission.

References

Aca, J., Mejía, R., Velandia, M., García, E., Galeano, N., Ahuett, H., et al. (2003). Integrated product development in virtual enterprises supported by Web-based applications. In L. M. Camarinha-Matos & H. Afsarmanesh (Eds.), *Process and foundations for virtual organizations* (pp. 361-368). Boston: Kluwer Academic Publishers.

Ávila, P., Putnik, G., & Cunha, M. (2002). Brokerage function in agile virtual enterprise integration" A literature review. In L. M. Camarinha-Matos (Ed.), *Collaborative business ecosystems and virtual enterprises* (pp. 65-72). Boston: Kluwer Academic Publishers.

Bichler, M., & Segev, A. (1999). A brokerage framework for Internet commerce. *Distributed and Parallel Databases, 7*(2), 133-148.

Boyson, S., Corsi, T., Dresner, M., & Harrington, L., (1999). *Logistics and the extended enterprise.* New York: John Wiley & Sons.

Boyson, S., Corsi, T., & Verbraeck, A. (2003). The e-supply chain portal: A core business model. *Transportation Research Part E, 39*(1), 175-192.

Bremer, C. F., Eversheim, W., Walz, M., & Molina Gutiérrez, A., (1999). Global virtual business: A systematic approach for exploiting business opportunities in dynamic markets. *International Journal of Agile Manufacturing, 2*(1), 1-11.

Bultje, R., & van Wijkt, J. (1998). Taxonomy of virtual organizations, based on definitions, characteristics and typology. *Virtual-Organization.Net - the Newsletter, 1 & 2*(3), 7-20.

Byrne, J. A., Brandt, R., & Port, O. (1993). The virtual corporation. *Business Week,* February 8, 98-102.

Caballero, D., Molina, A., & Bauernhansl, T. (2000). A methodology to evaluate enterprises to become members of virtual industry clusters. In L. M. Camarinha-Matos, H. Afsarmanesh, & Rabelo (Eds.), *E-business and virtual enterprises: Managing business-to-business cooperation* (pp. 443-454). Boston: Kluwer Academic Publishers.

Camarinha-Matos, L. M., & Afsarmanesh, H. (2003). Elements of a base VE infrastructure. *Computers in Industry, 51*(1), 139-163.

Camarinha-Matos, L. M., Afsarmanesh, H., Garita, C., et al. (1998). Towards an architecture for virtual enterprises. *Journal of Intelligent Manufacturing, 9*(2), 189-199.

Camp, R. C. (1989). Benchmarking: The search for industry best practices that lead to superior performance. *Quality Progress, 22*(2), 70-75.

Davidow, W. H., & Malone, M. S. (1992). *The virtual corporation.* New York: Harper Collins.

Evans, J., & King, V. (1999). Business-to-business marketing and the World Wide Web: Planning, managing, and assessing Web sites. *Industrial Marketing Management, 28*(1), 343-358.

Eversheim, W., Schuth, S., Bremer, C. F., & Molina, A. (1998). Globale virtuelle Unternehmen [Global Virtual Enterprise]. *ZWF, 93*, 62-64.

Ferreira, J. J. P. (2001). Dynamic forecast for master production planning with stock and capacity constraints EU Project: DAMASCOS. *International Workshop on Open Enterprise Solutions: Systems, Experiences, and Organizations.*

Filos, E., & Banahan, E. P. (2000). Will the organizations disappear? The challenges of the new economy and future perspectives. In *E-Business and Virtual Enterprises—Managing Business-to-Business Cooperation. Second IFIP Working Conference on Infrastructures for Virtual Organizations: Managing Cooperation in Virtual Organizations and Electronic Business towards Smart Organizations*, Florianópolis, Brazil, (pp. 3-20).

Flores, M., & Molina, A. (2000). Virtual industry clusters: Foundation to create virtual enterprises, in advanced in networked enterprises—virtual organizations. In L. M. Camarinha-Matos, H. Afsarmanesh, & Heinz-H. Erbe (Eds.), *Balanced automation and systems integration* (pp. 111-120). Boston: Kluwer Academic Publishers.

Fredenhall, L., & Hill, E. (2001). *Basics of supply chain management*. New York: St. Lucie Press.

Galeano, N. (2002). *Método para administrar las competencias clave en Clusters Virtuales Industriales* [Method to manage core competencies in Virtual Industry Clusters]. Thesis on Manufacturing Systems.

Goldman, S., Nagel, R., & Preiss, K. (1995). *Agile competitors and virtual organizations: Strategies for enriching the customer*. New York: Van Nostrand Reinhold.

Goranson, H. T., (2003). Architectural support for the advanced virtual enterprise. *Computers in Industry, 51*(1), 113-125.

Harbilas, C., Dragios, N., & Karetsos, G. (2002). A framework for broker assisted virtual enterprises. In L. M. Camarinha-Matos (Ed.), *Collabo-*

rative business ecosystems and virtual enterprises (pp. 73-80). Boston: Kluwer Academic Publishers.

Hassan, T. M., Carter, C., Hannus, M., & Hyvarinen, Y. (2001). eLEGAL: Defining a frame-work for legally admissible use of ICT in virtual enterprises. In K. Thoben, F. Weber, & K. Pawar (Eds.), *Proceedings of the 7th International Conference on Concurrent Enterprising: Engineering the knowledge economy through cooperation* (pp. 347-355). Bremen, Germany.

IBM (2001). Web Services Conceptual Architecture (WSCA 1.0). IBM Software Group.

Jacobson, G., &Hillark, J. (1986). *Xerox: American samurai.* New York: Macmillan.

Kanet, J., Faisst, W., & Mertens, P. (1999). Application of information technology to a virtual enterprise broker: The case of Bill Epstein. *International Journal of Production Economics,* 23-32.

Karch, K. M. (1992). Getting organizational buy-in for benchmarking: Environmental management at Weyerhaeuser. *National Productivity Review, Winter issue,* 13-22.

Katzy, B. R., & Schuh, G. (1999). The virtual enterprise. In A. Molina, J. M. Sanchez, & A. Kusiak (Eds.), *Handbook of life cycle engineering: Concepts, methods and tools.* New York: Chapmann & Hall.

Ketle, F. (1997). *Marketing* (Asia ed.). Beijing: Chinese People University Press.

Kovács, G., & Paganelli, P. (2003). A planning and management infrastructure for large, complex, distributed projects beyond ERP and SCM. *Computers in Industry, 51*(1), 165-183.

Lackenby, C., & Seddighi, H. (2002). A dynamic model of virtual organizations: Formation and development. In L. M. Camarinha-Matos (Ed.), *Collaborative business ecosystems and virtual enterprises* (pp. 37-44). Boston: Kluwer Academic Publishers.

Leymann, F., Roller, D., & Schmidt, M. -T. (2001). *Web services and business process management.* Retrieved November 3, 2001, from http://www.research.ibm.com/journal/sj/412/leymann.html

Main, J. (1992). How to steal the best ideas around. *Fortune, 19,* 102-106.

Mejia, R., Aca, J., Garcia, E., & Molina, A. (2002). E-services for virtual enterprise brokerage. In V. Marik, L. M. Camarinha-Matos, & H.

Afsarmanesh (Eds.), *Knowledge and technology integration in production and services* (pp. 141-148). Boston: Kluwer Academic Publishers.

Molina, A., Flores, M., & Caballero, D. M. (1999). Enquete sur les indicateurs de productivite et les meilleures methodes de l'industrie manufacturiere de Monterrey (Mexique) a travers une etude de Benchmarking. *Revue Francaise de Gestion Industrielle, 19*(3), 63-75.

Molina, A., Gonzalez, D., Galeano, N., Flores, M., & Caballero, D. (1999). *Experiences in enterprise modelling and integration in Mexico.* IEMC '99, International Enterprise Modeling Conference 99, June 14-16, Verdal, Norway.

Molina, A., Mejia, R., & Velandia, M. (2003). Core processes, methods and e-services to support Virtual Enterprise Brokerage. *International Journal of Networking and Virtual Organisations, 2*(1), 33-49.

Molina, A., Ponguta, S., Bremer, C., & Eversheim, W. (1998). Framework for global virtual business. *Agility & Global Competition, 2*(3), 56-69.

Panta, S., Sethia, R., & Bhandari, M. (2003). Making sense of the e-supply chain landscape: An implementation framework. *International Journal of Information Management, 23*(1), 201-221.

Perks, C., & Beveridge, T. (2003). *Guide to enterprise IT architecture.* Springer Ed..

Qin, Q., Xi, Z., & Chen, J. (2001). The innovation of e-marketing combination tactics. In W. Kim et al. (Eds.), Human.Society@Internet, LNCS 2105, 177-184.

Ramirez-Santaella, A., & Molina, A. (2004, April 5-7). *Multihead and Web Technology for the Implementation of a Manufacturing Execution System.* CD Preprints 11th IFAC Symposium on Information Control Problems in Manufacturing, Salvador-Bahia, Brazil.

Reddy, R., & Reddy, S. (2001). *Supply chains to virtual integration.* New York: McGraw-Hill.

RoundRose Associates, Ltd. (2004). *Smart organizations.* Retrieved March 24, 2004, from http://www.roundrose.com/smart.html

Roy, J., & Ramanujan, A. (2001). Understanding Web Services. IT Pro IEEE Computer Society.

Skonnard, A. (2002). The birth of Web services. *MSDN Magazine*

Trivedi, R. (2002, April). Web services architecture model RCG information technology. White paper.

Ullrich, D., Brockbank, W., & Yeung, A. (1989). Beyond belief: A benchmark for human resources. *Human Resources Management, 28*(3), 311-335.

W3C Working Group. (2004). Web Services Architecture Note 11. Retrieved March 4, 2004, from http://www.w3.org/TR/2004/NOTE-ws-arch-20040211/

Walleck, A. S., O'Halloran, J. D., & Leader, C. A. (1991). Benchmarking world-class performance. *The McKinsey Quarterly, 1*(1), 3-24.

Section III

Knowledge- and Human-Centered Technologies in Smart Organizations

Chapter IV

Knowledge Management in Smart Organizations

Shirley Chan, The University of Hong Kong, Hong Kong

Abstract

This chapter looks at the deployment of appropriate information and communication technologies in helping smart organizations to manage knowledge. Taking a management perspective, smart organizations can be regarded as those that can make smart strategic decisions and put into practice such managerial principles as value creation, continual learning, embracing uncertainty, and empowerment. Making good decisions would involve gathering and synthesizing the appropriate knowledge—knowledge about the market, products, suppliers, customers, competitors, and others. Different schools of knowledge management theories and the related technologies will be discussed. The author hopes that understanding the knowledge management technologies and related practices would assist researchers and practitioners in gaining some insights into managing the knowledge required for making smart decisions in organizations.

Introduction

The 21st century witnesses innovative organizational or work arrangements such as digital factory, virtual, or smart organizations. This kind of organization is like a network of independent production units working together, and is flexible and responsive to the challenges and uncertainty of the ever-changing business environment. The increasingly popular deployment of information and communication technologies facilitates these organizations in establishing unconventional work arrangements, networking with fellow co-workers or business partners, and seizing the opportunities as offered by the uncertainty in the current business climate.

Whether these organizations are forming innovative work arrangements, networking with stakeholders, or tapping into new opportunities, they are making strategic decisions in the Knowledge Age. The common thread running through such activities is *knowledge*—the ability to find, use, store, share, and retrieve the relevant organizational knowledge to make the right move and gain a competitive edge.

The term "smart organization" is used for organizations that are knowledge-driven, internetworked, dynamically adaptive to new organizational forms and practices, learning as well as agile in their ability to create and exploit the opportunities offered by the new economy (Filos & Banahan, 2000).

This chapter focuses on the knowledge sharing, network, and management in smart organizations. It will look closely at the management aspects of smart organizations and how knowledge management fits into the overall organizational management. Taking a management-oriented approach, the notion "smart organizations" appears to be originated from the book *Smart Organisation: Creating Value through Strategic R&D* (Matheson & Matheson, 1998). The authors consider "smart organizations" as those organizations that develop world-beating products on a continuing basis at prices that establish value leadership. They also regard "being smart" as making good decisions.

We are in the era of the Knowledge Economy. The basis of competition is "knowledge." Mastery of the relevant, crucial, and up-to-date knowledge would enable businesses and organizations to survive well and gain a competitive edge in this age. As stated above, "being smart" means making good decisions. Making good decisions would involve gathering and synthesizing the appropriate knowledge—knowledge about the market, products, suppliers, customers, competitors, regulatory environment, and other aspects. Such knowledge, if well utilized, would facilitate making good decisions so that the related organizations can develop world-beating products and services—befitting to be regarded as "smart organizations."

It would therefore be imperative to look at knowledge management in the context of smart organizations. This chapter will start off by introducing to the readers the general theoretical framework and the nine principles of smart organizations as indicated by Matheson and Matheson (1998), with the focus on how *knowledge* could be seen as a common thread running through these nine principles. This will be followed by an elaboration of the development and theory relating to knowledge management. It will then focus on how knowledge management helps realize the potential of smart organizations by implementing each of their nine principles. The relevant knowledge management technologies will also be discussed.

Background

Smart Organizations

There are different definitions of "smart organization." Some take a broader perspective, while others may be based on a narrower context. These definitions may have some differences according to the field of approach—such as management, information technology, or human resources. A management-oriented definition of "smart organization" can be found in Matheson and Matheson (1998), which will be elaborated immediately below. This chapter focuses on knowledge management in smart organizations as seen from this management-oriented approach.

The term "smart organization" appears to be originated from the title of the book *The Smart Organisation* by David Matheson and Jim Matheson (1998). They regard smart organizations as those companies that develop world-beating products and services and deliver them at prices that can establish value leadership. They are also of the opinion that "being smart" and "acting smart" would guarantee businesses to succeed in this fast-changing and increasingly competitive market.

They define "being smart" as making good strategic decisions. The authors consider "acting smart" as "the activity of effectively carrying out those decisions." Having said that, what are qualities expected of smart organizations? They conducted a benchmarking of best practices for strategic decision-making in R&D (research and development) in the early 1990s identified organizational characteristics determining whether companies would be successful in adopting best practices. They name these characteristics the "nine principles of the smart organisation" (Matheson & Matheson 2001). These nine principles about smart organizations as indicated in Matheson and Matheson (1998) are as follows:

- **Value Creating Culture:** The purpose of the organization should be maximizing the value created for customers and captured by the enterprise.
- **Creating Alternatives:** Choice means that several good alternatives should be created and that the related organization would choose the best one.
- **Continual Learning:** One of the main objectives of an organization should be learning continually about what would create value and how to deliver it.
- **Embracing Uncertainty:** People should endeavour to understand all sources of uncertainty and use that knowledge when making decisions.
- **Outside-in Strategic Principle:** The organization aims at understanding the dynamics of its industry and customers and using this perspective to frame and evaluate strategic decisions at all levels.
- **Systems Thinking:** It is advisable that the organization answers complex questions by thinking through cause-and-effect relationships in the context of the whole business and identifying leverage points, feedback loops, and key factors.
- **Open Information Flow:** It is crucial that in the organization, information is available to whomever wants it and that it is used to create value. Such flow of information should cross functional boundaries.
- **Alignment and Empowerment:** The people involved are empowered and trusted to pursue value creation and that the organization is guided by a shared understanding of its strategies for creating value.
- **Disciplined Decision Making:** This refers to a decision-making process that identifies strategic choices, involves the relevant people and information, and selects alternatives based on the highest value.

Smart organizations demonstrate most, if not all, of the above nine principles (Matheson & Matheson, 2001), which would enable them to make good strategic decisions. One could therefore argue that a common thread running through these nine principles is *knowledge*—gathering, utilizing, and understanding the relevant knowledge—knowledge about the industry, market, products, suppliers, customers, competitors, and other related issues to facilitate the organization in making good strategic decisions, rendering them to be considered as "smart." It is therefore imperative for smart organizations to understand the practices of knowledge management and the related enabling technologies so that they could make the best use of such practices and technologies to implement each of the nine principles.

Knowledge Management

Some wonder what is meant by "knowledge" and how it is different from information and data. Schoderbek, Schoderbek, and Kefalas (1985) regard "data" as " ... unstructured, uninformed facts so copiously given out by the computer. Data can be generated indefinitely; they can be stored, retrieved and updated and again filed...." They are also of the view that "information" refers to facts with meaning or evaluated data. One of the world's leading experts in knowledge management, Karl-Erik Sveiby, stated that information simply exists and is all waiting to be interpreted, to have meaning attached by people, and becomes knowledge at the moment of its human interpretation (Miller, 1999).

One of the management gurus, Earl (2001), is of the opinion that theoretical insights into how knowledge might be managed are available from several disciplines, including economics (Silberston, 1967), philosophy and epistemology (Kuhn, 1970), computer science (Hayes-Roth, Waterman, & Lenat, 1983) and sociology (Polanyi, 1958; Polanyi, 1966).

One of the world-renowned knowledge management gurus, Yogesh Malhotra, stated that the term "knowledge management" has been appearing in information systems literature over the past two decades, since the 1980s. This is because the knowledge management systems were traditionally included into the information systems research domain, with a heavy artificial intelligence and expert systems emphasis (Malhotra, 1999).

Throughout the 1990s, knowledge management as a discipline received more attention as managers in the post-Industrial Age had become more aware of the phenomenon that knowledge may be the most critical resource, rather than land, machines, or capital (Drucker, 1993). From a more practical perspective, knowledge management was considered to be central to product and process innovation and improvement, to executive decision-making, and to organizational adaptation and renewal (Earl, 2001). Different schools of thought regarding knowledge management emerged during the 1990s, which will be discussed immediately below.

IT Track and People Track

Karl-Erik Sveiby concurred with the above view that knowledge management has a technology-oriented origin. His observation was that during the 1990s and leading up to 2000, knowledge management specialists were from the "IT track." They were researchers and practitioners that have their education in computer and/or information science, and were involved in constructing information management systems, artificial intelligence, reengineering, and groupware. This "IT track" went through three rapid phases: (1) The first phase started around 1992 and focused on productivity issues such as "How can we use IT systems to prevent reinventing the wheel?" (2) The second phase had a customer focus addressing such concerns as "How can we leverage what we know about our customers to serve them better?" and (3) The third phase was sometime around 1999 to 2001, coinciding with the dot.com boom and there was much discussion about interactive IT Web pages, e-business, and online transactions (Sveiby, 2001).

During the knowledge management evolution along this IT track, there was the emergence of various schools of thought focusing more on the non-technical aspects of knowledge management, or one could name that as the "people track" (Sveiby, 2001)—that knowledge is an organizational resource and that people play a key role in the process of creating, utilizing, and managing knowledge. One of these is the "intellectual capital" school of thought led by Leif Edvinsson. He is probably the world's first corporate director in intellectual capital. In his capacity as the intellectual capital director of Skandia, the Swedish insurance and financial services company, he engineered the publishing of the world's first intellectual capital annual report (Anonymous, 1997). He advocated that such intangible assets as intellectual capital (like an employee with a PhD) could be measured, and the resulting value could be included in the annual financial statements (Edvinsson & Malone, 1997).

Advocates of the "people track" of knowledge management tied in knowledge management with business strategy and generating value. Thomas Davenport and Larry Prusak drew on their work with more than 30 knowledge-intensive organizations and examined how different kinds of companies can effectively understand, analyse, measure, and manage their intellectual assets, turning corporate knowledge into market value (Davenport & Prusak, 1997). Two Japanese experts, Nonaka and Takeuchi, offered a refreshing approach to knowledge management when they published their book *The Knowledge-Creating Company: How Japanese Companies Create the Dynamics of Innovation* in 1995 (Nonaka & Takeuchi, 1995). They proposed a SECI model for understanding the organizational knowledge creation process, which involves four stages: socialization, externalization, combination, and internalization (Nonaka & Takeuchi, 1995; Ursin, 2000). They further commented that the traditional Western view, emphasizing the organizational importance of knowledge that is codified and stored through such technical means as the databases, should be broadened to include *tacit* knowledge, which is knowledge based on experiences, insights, hunches, and perspectives, and often remains in human brains. Dave Snowden advocates that we should grow out of managing knowledge as a *thing,* but rather as a *flow*—which means we should focus more on narrative and context when sharing or managing knowledge, rather than just the content. This emphasis on narrative suggests that we can always know more than we can tell and we will always tell more than we

Table 1. Knowledge management schools of thought: IT track and people track

IT or People Track?	Development Phase or Focus	Main Issue
IT Track	First Phase—Productivity Focus	How can we use IT systems to prevent reinventing the wheel?
	Second Phase—Customer Service Focus	How can we leverage what we know about our customers to serve them better?
	Third Phase—Dot.com Boom	Discussion about interactive web pages, e-business, and online transactions.
People Track	Market Value and Competitive Advantage	Organizations can understand, analyse, measure, and manage their intellectual assets and turn their organizational knowledge into market value.
	Knowledge Creation	SECI Model—Organisational knowledge creation process involves four stages: socialization, externalization, combination, and internalization.
	Narrative and Context	We manage knowledge as a flow and not as a thing.

can write, because the process of writing reflective knowledge is time consuming and involves loss of control over its subsequent use (Snowden, 2002).

Technocratic, Economic, and Behavioural Approaches

The "IT track" and "people track" approaches as proposed by Sveiby (2001) could be one way of viewing the developments of the knowledge management field. Earl (2001) suggested an alternative model of viewing such developments—three approaches consisting of seven schools. The first three schools are labelled "technocratic," as they are based on the premise that information technology supports and, to various degrees, conditions employees (or knowledge workers) in their daily tasks. The first school is the systems school, whose fundamental idea is to capture specialist knowledge in knowledge bases (from conventional databases through CD-ROMs to expert systems), which other specialist or "qualified" people can access. The second school is the cartographic school, which is concerned with mapping organizational knowledge and aiming at recording and disclosing who in the organization knows what by building knowledge directories to ensure that knowledgeable people in the organization are accessible to others for advice and consultation. The third school is the process school, which is a derivative or outgrowth of business process engineering. There are two ideas driving this school:firstly, performance of business processes can be enhanced by providing operating personnel with knowledge relevant to their tasks, and secondly, management processes are more knowledge-intensive than business processes (Earl, 1994,2001).

While the first approach emphasizes the "technocratic" aspect, the second approach is "economic" in nature because it has a business orientation explicitly creating revenue streams from the exploitation of knowledge and intellectual capital. It is perhaps what Davenport, De Long, and Beers (1998) described as "managing knowledge as an asset," where knowledge or intellectual assets consist of patents, trademarks, copyrights, and know-how (Earl, 2001).

The third approach is "behavioral" in nature, as it seeks to stimulate and orchestrate managers or managements to be proactive in creating, sharing, and using knowledge as a resource, and there are three schools of thought under this head. The organizational school emphasizes using organizational structures or networks to share or pool knowledge. They are often described as "knowledge communities," which are groups of people with common interests, problems, or experiences. These communities are designed and maintained for a business objective, and could be either interorganizational or intraorganizational in terms of arrangement. Although such communities may be supported by technology, this school is still under the "behavioural" approach because the essential feature of these communities is that they exchange and share knowledge interactively and often in nonroutine, personal, and unstructured ways, as an interdependent network. Another school is the spatial school, which centers on the use of space—or spatial design—to facilitate knowledge exchange or sharing. Typical examples include the metaphors quoted in modern management journalism—such as the water cooler as the meeting place, the kitchen as a "knowledge café" or the open-plan office as a "knowledge building." These are essential "spaces" for knowledge transfer within modern organizations, as most people are social beings who like human contact and often prefer conversation to documents or information systems. Tacit knowledge is most likely to be discovered and exchanged when people socialize within such spaces—and an alternative label for this school could be social school, because the rationale is to encourage socialization as a means of knowledge exchange. Finally, the strategic school is primarily concerned with raising consciousness about the value creation possibilities available from recognizing knowledge as a resource, and this explains why some corporate mission and purpose statements embrace knowledge as an important item on the organizational agenda (Earl, 2001).

Table 2. Different schools of thought regarding knowledge management

Approach	School of Thought	Main Issue
Technocratic	Systems	Capture specialist knowledge in knowledge bases (e.g., databases) which specialist people can access.
	Cartographic	Map organizational knowledge by building knowledge directories so that others in the organization can consult them.
	Process	Business processes can be enhanced by providing operating personnel with relevant knowledge. Management processes are more knowledge-intensive than business processes.
Economic	Economic	It has a business orientation explicitly creating value streams from the exploitation of knowledge.
Behavioural	Organisational	It emphasizes using organizational structures or networks to share or pool knowledge such as "knowledge communities."
	Spatial	It centers on the use of space to facilitate knowledge exchange or sharing.
	Strategic	It is concerned with raising consciousness about the value creation possibilities available from recognizing knowledge as a resource.

Knowledge Management Practices and Technologies: Implementing the Nine Principles

As stated above, there are generally nine principles characterizing an organization to be a smart organization and that one could argue that knowledge is the common thread running through these nine principles, as smart organizations need to make good strategic decisions in the knowledge age. Below will be a discussion on how common knowledge management practices and technologies would help implement or realize each of the nine principles, rendering an organization smart. During the discussion, current knowledge management theories and schools of thought will be included as appropriate.

Value Creating Culture

It is crucial for a smart organization to aim always for value creation. Within such organization, it would be ideal if every person saw his or her final job as creating the greatest value. There could be changes or exceptions to rules if these rules stand in the way of creating value. In respect to organizational culture, it aims at rewarding those who work to create more value and encourages different functions to collaborate to create value. In case there are disagreements within the organization, the policy is that they should be resolved in favor of attaining the greatest value (Matheson & Matheson, 1998).

The conventional way of understanding "value creation" in the past two decades has been influenced by the notion of industrial logic, in which value is added in sequential stages based on the premise that each part can be optimized individually and therefore contributing to the overall organizational value creation (Ramirez, 1999; Roberts, 2000; Skoog, 2003). During the past two decades, nonlinear models have been challenging this traditional value creation framework (Heskett, Jones, Loveman, Sasser, & Schlesinger, 1994; Simons, 1995; Ramirez, 1999; Roberts, 2000; Skoog, 2003). These new perspectives emphasize the role of different actors (stakeholders) as coproducers or creators of organizational value. They appear to stress the significance of combinations and connectivity among the related actors or stakeholders and assert that a significant part of the value creation cannot be expressed in mere monetary terms (Heskett et al., 1994; Simons, 1995; Ramirez, 1999; Roberts, 2000; Skoog, 2003) because "value," or the currency in the current knowledge economy is "knowledge," which is an intangible asset not normally measured in terms of money.

Our knowledge economy is characterized in terms of the nature of its products and services being knowledge-based. Value creation in the knowledge age is therefore very much related to strategically leveraging an organization's intellectual or knowledge capital. It is important for an organization to have the capacity to coordinate, orchestrate, and deploy its knowledge resources toward creating value in the pursuit of its future vision. In fact, the nature and rationale of intellectual capital is about creating value through developing and deploying knowledge-based competitive advantages faster than the competitors or the pace of change in the industry (Rastogi, 2003). Knowledge and its exploitation for creating value are the conjoint outcomes of organizational learning (Senge, 1990; Rastogi, 2003). During this process, an organization may have to meet many challenges and exploit various opportunities in its quest for value creation (Rastogi, 2003). Effectively managing knowledge would help organizations to overcome such challenges and make the best use of such opportunitics so as to achieve the ultimate goal of creating value.

Extranets

Our current business environment is becoming more competitive, with customers having a wide variety of consumption choices, each alleging to be offering high-quality products or services with a relatively lower price. In this context, "value'" can be seen in the context of the relationship between supplier offerings and customer purchases by identifying how the supplier fulfils the customer's needs (Band, 2000; Porter, 1998; Clarke, 2001). Mastering the knowledge about the customer's needs or requirements would be of utmost importance in creating value for a smart organization.

The knowledge management technology of extranets could help organizations better understand customers' needs and deliver a better service and thereby create value. Extranets are secure Internet protocol-based networks linking the information infrastructures of various extranet participants—in the business context, usually one or more of the following parties: suppliers, vendors, customers, and business partners (Bushko & Raynor, 2001). Extranets are a result of the evolution of electronic document exchange (EDI) technology, which has been used for many years to connect organizations together for supply chain integration. EDI used the automated exchange of simple, highly structured electronic forms over private networks, whereas extranets dramatically extend the benefits of interorganizational integration through the intertwining of EDI with Internet technologies. Since the Internet is a public and international network based on the Internet protocol (IP) and related standards, it provides a standard of interconnecting networks so that any system can communicate with any other system. Hence, extranets have the advantage of allowing various organizations having different hardware and software to communicate given extranets' use of the Internet's open standards (Yen & Chou, 1999, 2000).

One of the common purposes of extranets is to allow various organizations to share their information resources and knowledge storage, facilitate their knowledge flow, and work with their suppliers, customers, distributors, and other businesses to reach a common goal. The knowledge flow could be in relation to the customers' knowledge about the vendors' products and services, or business partners sharing their knowledge about certain target markets with the view of setting up joint ventures there. The essence of the extranets, from a knowledge management perspective, is allowing each of the extranet partners to access certain permitted areas of their business partners' networks in order to gain access to the requisite knowledge.

Such knowledge management could add value to the service delivered to the clients. For instance, extranets linking an organization with its customers or clients are instrumental in strengthening client relationships and creating value for clients. An example would be Deloitte Consulting's UK practice developing a highly technologically sophisticated extranet to link up a client firm in respect of a major strategy project. It uses the extranet to communicate sensitive information to a select audience (in this case, the client firm), keep the clients and other stakeholders informed of the strategy formulation process, and allow the clients access to the detailed information about the project as the project goes on (Bushko & Raynor, 2001). The added "value" as exemplified in this case is enhanced client service quality—in the sense that in addition to Deloitte providing high-quality consulting service, the client firm can actually get access to the relevant information sources at Deloitte to keep track of the internal workings of the project and thereby facilitate consultant/client relationship.

Customer Relationship Management Systems

While extranets connecting an organization's networks to its customers may have the potential of enhancing the service quality, the technology of customer relationship management systems would help manage knowledge about customers' needs and customer relationships and enable the organization to create value. Customer relationship management is about using information technology in implementing relationship marketing strategies. Berry (1983) first introduced the term "relationship marketing," which is concerned with how organizations manage and improve their relationships with customers for long-term profitability (Ryals & Payne, 2001). He defined relationship marketing as "attracting, maintaining, and ... in multi-service organizations ... enhancing customer relationships" (Berry, 1983). Since then, the field of relationship marketing has generated much interest and attention (Sheth, 2000). The longer the customer relationship lasts, the more profitable customers are shown to be, and therefore the focus of customer relationship management is on the lifetime value of the customer, rather than the profitability in any single period (Reichheld, 1996). Relationship marketing advocates view the customer relationship as an asset that can be managed and that requires investment. Technology can help organizations manage the information that they need to understand customers, so that appropriate relationship marketing strategies can be formulated (Ryals & Payne, 2001). This is where customer relationship management systems play a role.

Kutner and Cripps (1997) contend that customer relationship management is based on four premises: (1) manage customers as important assets; (2) customer profitability varies, and not all customers are equally desirable; (3) customers vary in their needs, preferences, buying behaviour, and price sensitivity; and (4) by understanding customers drivers and profitability, organizations can tailor the products and services they offer in order to maximize the overall value of their customer portfolio (Ryals & Payne, 2001). Kotler (1990) is of the view that organizations should make the best use of technology to manage their customer relationships. They could build and use customer databases to keep track of what customers are buying and what they are interested in, therefore utilizing such information to serve customers better (as quoted in Caruso, 1992; Ryals & Payne, 2001).

Such information can be used to strengthen the relationship with the customer and increase customer value over time (Gronroos, 1997). For instance, First Direct bank uses the data from a previous transaction in a proactive way in order to strengthen the relationship by personalizing a later transaction. The customer, who wishes to use his First Direct bank card in an American ski resort, asks for information about automated teller machines there. In a subsequent transaction, a different call center operator uses this information to check whether this customer had a satisfactory experience. External data sources can also be added to the store of knowledge about customers. Market research results or information from other databases can also be added to this customer knowledge store (Ryals & Payne, 2001).

As mentioned above, creating value requires knowledge and understanding of the customers, their needs and requirements, and other related factors. These factors could be the market, competitors, and regulatory environment, which could impact the customers' needs. Such an understanding would help the organization evaluate alternatives and the business environment, find the appropriate strategic position, and make a wise decision rendering the organization to add value to the customers.

Creating Alternatives and Disciplined Decision-Making

Disciplined decision-making is a process identifying strategic choices, engaging the right information and people and selecting alternatives based on the highest value. People working in smart organizations should understand the nature, importance, and process of decision quality and initiate the appropriate process in addressing the strategic decisions. Part of the culture of the smart organizations is that a quality decision is applied to every important decision, including portfolio, technology, and R&D project strategy (Matheson & Matheson, 1998).

As disciplined decision-making involves identifying strategic choices, there would be a preceding step, which is creating alternatives. The notion of creating alternatives is about looking for high-value alternatives that are desirable, safe, and rewarding. It also requires committing to evaluating alternatives honestly and without prejudice, if possible. Generating alternatives would pave the way and allow for more choices when it comes to the stage of identifying strategic choices. Choice means creating various good alternatives and selecting the best one as appropriate for the situation (Matheson & Matheson, 1998).

The processes of alternative creation and disciplined decision-making—generating, identifying, and evaluating options and selecting the most appropriate one—would be facilitated by knowledge sharing and retrieval. This is because one that is well-equipped with the requisite knowledge would be in a position to generate the relevant alternatives and make an informed decision. Sharing knowledge among people concerned and retrieving knowledge from databases or other knowledge storage places would help maximize the knowledge base on which an organization or each individual within that organization can create alternatives and make a good strategic decision. The knowledge base could be in relation to the knowledge about the industry, markets, customers, suppliers, competitors, and other stakeholders.

Knowledge sharing and knowledge retrieval can be enabled by the related knowledge management technologies. Part of the organizational knowledge generation process consists of sharing knowledge in the first place and surfacing current knowledge and assumptions, making it available for critical scrutiny (Despres & Chauvel, 2000). Knowledge can also be retrieved from where it is stored for understanding, analysis, and utilization. The knowledge management technologies of intelligent agents could enable the knowledge sharing and retrieval processes, providing the knowledge base for generating alternatives and engaging in disciplined decision-making.

Intelligent Agents

An "agent" in the legal sense is empowered to act on behalf of another (Feldman & Yu, 1999). An "intelligent agent" is one that can learn the behavioural patterns or the rules concerning certain actions and transactions, and then act accordingly on behalf of its "boss" (Feldman & Yu, 1999) or user. While there would be various definitions of "intelligent agents," most current researchers agree that agents have the following characteristics: autonomy, adaptiveness, collaborative behaviour, and mobility (Feldmand & Yu, 1999).

Agents have knowledge retrieval, profiling, and filtering capacities. They can search and retrieve information brokers or document managers (Stenmark, 2003). They often play a collaborative roleproviding information and expertise on a specific topic by drawing on relevant information from other information agents. The architecture of an agent is that each agent contains a domain model (providing descriptions of the classes of objects in the domain, relationships between these classes, and other domain-specific information) and information resource models (providing descriptions of both the contents of the information sources and the relationship between those models and the domain model) (Knoblock, Arens, & Hsu, 1994).

The system uses these mappings for transforming a domain-model query into a set of queries to the appropriate information sources. Information retrieval query processing requires developing a plan for obtaining the data. This includes selecting the information sources to provide the data, the processing operations, the sites where the operations will be performed, and the order of performing (Ambite & Knoblock, 1997). The organization of agents needs a common communication language and protocol to interact and collaborate. A common content language is the Loom knowledge representation language (MacGregor, 1990), which is a language for representing hierarchies of classes and relations, as well as efficient mechanisms for classifying instances of classes and reasoning about descriptions of object classes (Knoblock et al., 1994). The knowledge query and manipulation language (KQML) is a common protocol to organize the dialogue between agents (Ambite & Knoblock, 1997), as explained above.

Knowledge Interchange Format

While the intelligent agents are looking for the relevant information, they may have to communicate or establish links with databases or knowledge sources residing in disparate computer systems. Knowledge interchange format is a language designed for use in the interchange of knowledge among disparate systems, which could be those created by different programmers, at different times or in different languages. The purpose of knowledge interchange format is quite analogous to that of Postscript, which is commonly used by text and graphics formatting systems in communicating information about documents to printers. Postscript is a programmer-readable representation facilitating the independent development of formatting programmes and printers. While knowledge interchange format is not as efficient as a specialized representation for knowledge and not as perspicuous as a specialized display, it is a programmer-readable language capable of facilitating the independent development of knowledge-manipulation programmes (Genesereth, 2004).

The features essential to the design of knowledge interchange format include: (1) The language has declarative semantics so that it is possible to understand the meaning of an expression in a language without appeal to an interpreter for manipulating that expression; (2) The language is logically comprehensive in the sense that it provides for the expression of arbitrary logical sentences; and (3) The language provides for the representation of knowledge about knowledge, allowing the user to make knowledge representation decisions explicit and permitting the user to introduce new knowledge representation constructs without changing the language (Genesereth, 2004).

As explained above, intelligent agents with information retrieval and filtering abilities have a major characteristic distinguishing them from search engines: its proactive nature. Search engines such as Yahoo! and Google are inherently reactive (an information seeker has some query in mind, puts in some keywords, and the programme reacts to such query by displaying the search results). Retrieval and filtering agents are proactive in the sense that they keep watching a user's environment (usually a computational environment such as e-mail or a Web page that a user is reading) and present information to the user without requiring any continual action of the part of the user (Rhodes & Maes, 2000).

Creating alternatives and engaging in disciplined decision-making would need the decision-makers to align the organizational goals, because any alternative created or decision made without the possibility of ultimately attaining organizational goals would not lead to success for the organizations concerned. Such alternative creation and disciplined decision-making processes could also employ systems thinking involving the consideration of various factors or possibilities before reaching a final decision.

Alignment, Empowerment, and Systems Thinking

Smart organizations encourage participation in the decision-making process to achieve alignment of goals and the understanding required to make empowerment effective. The organization is guided through a shared sense of understanding of its strategies for creating value, and that people are empowered and trusted in pursuing value creation. Employees in the organization should feel empowered to act and take on the responsibility of acting, as well as for maintaining a shared sense of purpose. The organizational culture should be encouraging strategic decisions to be through a participative process and horizontal and vertical dialogue in realigning the organization through various periods of change (Matheson & Matheson, 1998).

While the organization is going through the decision-making process to achieve alignment of goals, it would be advisable that it deploys systems thinking—considering various factors before making a strategic decision. Systems thinking involves answering complex questions by thinking through cause-and-effect relationships from the perspective of the whole business and identifying leverage points, feedback loops, and key factors (Matheson & Matheson, 1998). The system approach can be described as a method to assemble and organize information, knowledge, and activities to attain greater efficiency (Vallee, 2003).

Systems thinking also requires an agile mind and a keen appreciation of how various parts of the world (the system) are linked together. Employees in an organization promoting systems thinking are expected to understand how their jobs and their actions are part of a larger system, and work to incorporate multiple and whole-system approaches into their thinking. Decision-makers in this kind of organization expect people to think through the full implications of their proposals and bring multiple perspectives to each important choice (Matheson & Matheson, 1998).

Organizational Decision Support Systems

The knowledge management technology of organizational decision support systems may be able to help organizations utilize the collective knowledge of various individuals within the organization, encouraging participation in decision-making and deploying systems thinking by considering various factors before making a strategic decision. Decision support technologies, including decision support systems, are computer-based tools developed to provide managers with the relevant information about internal operations and its business environment, together with experts' knowledge and models to facilitate their decision-making (Cascante, Plaisent, Bernard, & Maguiraga, 2002). Given that alignment of goals, one of the characteristics of smart organizations, involves encouragement of participation in the decision-making of the organization, organizational decision support systems that aim to support organizational decision-making that cuts across functional boundaries (Kim, Graves, Burns, & Myung, 2000) appear to be most suitable among various other kinds of decision support technologies.

The key notions in the organizational decision support systems frameworks are distributed problem-solving by human and machine knowledge processes, communication among these problem-solvers or decision-makers, and coordination of interrelated problem solving or decision-making efforts in the interest of solving an overall decision problem (Holsapple & Winston, 1996; Kim et al., 2000). Organizational decision support systems refer to a collection of individual decision support systems that communicate with each other to support collective organizational decision-making. Each constituent decision support system depends on the while but capable of supporting local decision-making. While the objective of an individual decision support system is to improve the performance of individual decision-making, the goal of organizational decision support systems is to improve the performance of collective organizational decision-making (Kim et al.) by having the systems take into account the collective knowledge sources and organizational goals, and assisting the decision-makers to reach an optimum decision compatible with such goals.

While there may be variations from one type of organizational decision support system to another, a typical organizational decision support system architecture consists of four components: (1) organizational participants; (2) language subsystem; (3) messaging subsystem; and (4) public resource management subsystem (Kim et al., 2000). The organizational participants are either human or machine processors, which communicate with each other to solve overall organizational problems (Kim et al.). The Language Subsystem provides the user with the interface to accept requests from participant nodes and to display the results of the request (Holsapple & Whinston, 1996; Burns, Rathwell, & Thomas, 1987). The Messaging Subsystem involves communication among participant nodes, such as notification of actions of a participant node to others (Holsapple & Whinston, 1996; Swanston & Zmud, 1989; Burns et al., 1987). The Public Resource Management System manages the public resources, including data and models such as a directory of the participant nodes' names and operators' names that each decision support system wants to be available to (Burns et al., 1987; Kim et al.).

One example of how an organizational decision support system assists with collective decision-making is an organizational decision-making system helping a television station to forecast television viewership, which helps formulate its marketing and planning strategies. Factors influencing television viewership are closely related to the relevant knowledge in respect of sociocultural factors affecting viewership such as different lifestyles, the viewers' ages and program preferences, and weather conditions. Forecasts for the television viewership done manually relied on the experience of the people doing such forecasts, and producing such results was a very long process based on individual subjective assessments. The results were inaccurate because there was no quest for supporting evidence and the person who did the forecasting might have had all types of prejudices and limitations that usually appear in judgmental forecasting (Patelis, Metaxiotis, Nikolpoulos, & Assimakopoulos, 2003).

An organizational decision support system called FORTV was proposed, which can help decision-makers identify various factors affecting television viewership and the competitive environment, such as total viewership, market share of each TV channel, viewership of each TV station, the program's viewership and the commercials' viewership, and to have such relevant body of knowledge incorporated into the analysis process. The system can perform the function like a knowledgeable human being by forecasting television viewership in the future to help the organization plan its programs and devise its strategies (Patelis et al., 2003).

Empowering individuals within an organization to align organizational goals and deploy systems thinking in reaching a strategic decision would also require taking an outside-in perspective—appreciating the impact of the business environmental circumstances and the uncertainty they often represent, and evaluate how such external circumstances and uncertainty could affect the organization and its success.

Outside-in Strategic Perspective and Embracing Uncertainty

The essence of an outside-in strategic perspective is to see and act on the "big picture," appreciate its importance for strategy, and distance oneself from his or her personal circumstances. The organization would need to understand the dynamics of the industry and customers and use this perspective to frame and evaluate strategic decisions at all levels. The organizational culture should encourage actively seeking information about what is outside of the organization and use it in decision-making, and view inside-out frameworks with suspicion (Matheson & Matheson, 1998).

If gaining knowledge about the world outside of the organization is crucial for strategic decision-making, the organization would also need to have the qualities to embrace uncertainty given that the environment in which businesses operate these days is characterized by uncertainty and the only constant is change. Perceived environmental uncertainty exists when decision-makers are not confident about understanding what the major events or trends are in an environment, or that they feel unable to accurately assign probabilities to the likelihood that particular events and/or changes will occur (Miliken, 1987). Scanning the environment could also be a difficult organizational process because the environment is complex (Cyert & March, 1963) and becoming more uncertain.

People of a smart organization should seek to understand all sources of uncertainty and apply such knowledge when making decisions (Matheson &

Matheson, 1998), as strategic action depends on perceptions and interpretations of the environment (Schneider & De Meyer, 1991), and the environment is a major source of uncertainty for managers (Elenkov, 1997). It has been said that strategies are formulated in light of the perceived environmental conditions (such as uncertainties about the competitors, customers, and the environment) as well as internal capabilities (Parnell, Lester, & Menefee, 2000).

Research has identified multiple dimensions of the environment, such as dynamism, complexity, munificence, and uncertainty (Dess & Beard, 1984; Sharfman & Dean, 1991). Uncertainty has been defined as a combination of such perceived dynamism and complexity as they are held by the managers (Duncan, 1972; Koberg, 1987). Members of the organization are recommended to recognize that amid such uncertainty, decisions can be controlled but outcomes cannot. They need to understand uncertainty within their area of expertise, communicate such uncertainty accurately, and articulate it in terms of possibilities and probabilities with a realistic understanding of what they can influence and what may be beyond their control. The related organizational culture is one that promotes treating uncertain information as ranges or probability distributions. Forecasts are never turned into commitments because of the awareness of the uncertainty factor in the forecasts, and the decision-making process requires explicit consideration for risk and return (Matheson & Matheson, 1998).

Taking an outside-in strategic perspective requires knowledge about the external environment (such as knowledge regarding the customers, suppliers, and competitors) and embracing uncertainty calls for utilizing such knowledge to frame an analytical framework from which solutions and ideas in managing such uncertainty come up. Knowledge about customers, supplier, competitors, and industry may be gathered from various information sources within the organization, such as customer records and customer service survey results, market segmentation analysis, suppliers' order records, competitors' intelligence reports, and industry forecast. These various pieces of information may be scattered in different sections of the organization—for example, in customer service, marketing, logistics, and strategy planning departments respectively. The organization cannot have a complete picture of the external environment if it possesses knowledge of only one or two aspects of the environment—for instance, just the competitors and the industry. It needs knowledge of *all* relevant aspects of the environment, and this is where the knowledge management technology of enterprise informal portals comes in.

Enterprise Information Portals

An enterprise information portal can be defined as a *single* point of access for the pooling, organizing, interacting, and distributing of organizational knowledge (Aneja, Brooksby, & Rowan, 2000; Kendler, 2000; Schroeder, 2000). Enterprise portals have quite complex structures and features, but their basic functions and elements are relatively easy to define (Raol, Koong, Liu, & Yu, 2002). Firstly, from an operational point of view, the strength of enterprise portals lies in its ability to provide Web-based access to the organizational information, applications, and processes (Raol et al., 2002). Secondly, from a functional perspective, enterprise portals leverage existing information systems, data stores, networks, workstations, servers, and applications, as well as other knowledge bases, to give each individual within the organization immediate access to an invaluable set of organizational data anytime and anywhere (Kendler, 2000; White, 2000). As mentioned above, it is important for the organization to have knowledge about *all* relevant aspects of the environment when it takes an outside-in strategic perspective, and this function of enterprise portals being able to gather an integrated set of relevant information would help the organization in gathering the related knowledge in taking such perspective.

Enterprise portals could help an organization tap into not only its internal knowledge bases, such as the customer records, market segmentation reports, and industry forecasts, but also external sources about the environment, and therefore assist with the organization to take an outside-in perspective. In addition to ubiquity and ease of use of the Web browser interface, one of the important features of the enterprise portals is the availability of innumerable new data sources on the public Internet, in addition to the data sources across the organization (Kim, Chaudhury, & Rao, 2002). This has to do with the two-layered architecture of many enterprise portals, as explained below.

Aneja et al. (2000) comes up with a generic framework of an enterprise portal showing some of the major applications, entities, capabilities, tools, and their relationships, which was reproduced in Raol et al. (2002). The enterprise portal framework essentially contains two primary layers. At the core of any enterprise portal framework are the applications that it purports to support, which can range from unit-specific to organizational-wide capabilities, staff to administrative support functions, and individual to system-wide inquiries. Examples of such applications are office documents, decision support systems analysis/reporting, business content such as marketing and human resources information, personal or group Web sites, as well as collaboration facilities like e-mail and calendar

(Aneja et al., 2000). The second layer consists of various Web-based drivers (Raol et al., 2000) leading to external information sources such as Web sites, news feeds, stock and weather information, or travel reservations (Aneja et al., 2000). These drivers are the means for the openness and easy access capabilities to the disparate databases and reports generated. Some of the core enterprise portal software functions include customization and personalization, collaboration and community, content management, ease of use, dynamism, and security (Raol et al., 2000).

The current political, business, and social environment is characterized by constant change. Individuals and organizations that are able to rise to the challenge of taking an outside-in perspective and appreciating the uncertainty of such an environment are those that engage themselves in learning about new events, circumstances, ideas, and people. Open information flow as a way for individuals within the organizations to gather new knowledge, as well as continual learning, are of crucial importance in this context.

Open Information Flow and Continual Learning

The purpose of having an open information flow in an organization is to "inform and be informed." The flow of information crosses functional boundaries, with virtually all information available to whomever that wants it. Information is routinely captured, packaged, shared, and applied, and used in various ways to create value. People in a smart organization feel safe sharing what they know, and feel obliged to contribute to information sharing systems and are excited about learning and teaching. The prevailing organizational culture should support an ethic of both "giving and getting" in relation to information sharing, and this would hopefully discourage information hoarding (Matheson & Matheson, 1998). The organizational culture drives the overall value system, providing norms for information sharing and reaching a consensus on its meaning (Sinkula, 1994). Such information sharing is essential for the organization to learn continually, so as to address and overcome various challenges arising from operating in this increasingly complex and ever-changing business environment.

A smart organization should have the purpose of learning continually about what creates value and how to deliver it (Matheson & Matheson, 1998). Learning would enrich the intellectual capital of both individuals and organizations, and such capital is a significant resource in generating value in today's knowledge economy. For a number of years, some of the world's most proactive observers of societal change have predicted the emergence of a new economy in which intellectual prowess, not machine capability as valued in the industrial economy, would be the critical resource (Graham, 1996).

It would be ideal if people in a smart organization were not only excited about learning and growing, but also willing to accept constructive comments and new ideas and apply the same to themselves and their organization (Matheson & Matheson, 1998). All individuals within the organization are continually engaged in learning, helping each other to learn, and sharing their learning (Lawler, 1988) to the extent that it would become a "learning organisation." A learning organization is an organization that purposefully adopts structures and strategies that encourage learning (Dodgson, 1993). When such an organization emphasizes *continual* learning, the organizational culture would be viewing *change* as important, emerging, and profitable, and as something routine leading to improvements (Matheson & Matheson, 1998). The organization should also adopt a critical attitude questioning organizational myths and assumptions, and even welcoming "bad news," which will be used to initiate improvements (Matheson & Matheson, 1998).

Learning occurs when organizations seek not just to synthesize but also institutionalize people's intellectual capital and learning, their memories, cultures, routines, and core competencies. Though people may come and go in an organization, its memories preserve the individuals' behaviour, norms, and values as accumulated over time, and gradually build an organizational structure that will become the repository for lessons learned as the organization addresses and solves its problems on its way. As the members of the organization leave and new ones join, it is crucial that the knowledge and competence of the former staff can be transferred to the new ones across generations of learning (O'Keeffe, 2002). The related knowledge management practices and technologies could facilitate this cross-generational learning process.

Communities of Practice

Knowledge management advocates support the idea of forming communities of practice in facilitating organizational learning process. The idea of a community of practice was developed by Lave and Wenger (1990) as a theory for practice-based learning in which one could undertake "legitimate participation" to serve a kind of apprenticeship with a group of "insiders" in an organization, organizations consisting of a range of different disciplinary groups or collectives, each charged with specific areas of responsibility. Wenger and Snyder (2000) later depicted a community of practice as a "group of people informally bound together by shared expertise and passion for a joint enterprise," with members inevitably sharing knowledge in order to solve problems in their organization (Russell, Calvey, & Banks, 2003).

Wenger (1998) incorporated both informational and interactive elements of knowledge into his community of practice theory through the concepts of reification and participation. Reification is "the process of giving form to our experience by producing objects [including symbols and texts] that congeal this experience into 'thingness'" (Wenger, 1998, p. 58). Reification represents the concreteness apparent in the informational view of knowledge. Participation gives the active and social element of knowledge through engagement in the activity. Reification and participation together form a duality interacting in the process of knowing (Iverson & McPhee, 2002).

Wenger (1998) further identified three characteristics of communities of practice: (1) mutual engagement, (2) negotiation of a joint enterprise, and (3) a shared repertoire. Mutual engagement involves interaction with other members within the community of practice during which members are motivated to negotiate their practices and the meanings of their actions. By being mutually engaged with one another, knowledge is shared and enacted. Members can offer to each other insights, adopt others' practices, critique practices, and share frustrations, and hopefully the members will learn from each other through open flow of information during such interaction. Negotiation of a joint enterprise gives a sense of purpose and coherence to the community of practice. It can be said to be the common purpose binding people together and providing a unifying goal and coherence for their actions. Wenger states that this negotiation process creates more than "just a stated goal but creates among participants relations of mutual accountability that become an integral part of the practice (Wenger, p. 76). The third characteristic is shared repertoire, which refers to the community's set of resources for negotiating meaning. Knowing the shared repertoire such as jargon, stories, and other forms of a stock of understood information and techniques as utilized by the community members can be a proof of membership (Iverson & McPhee, 2002).

Intranets

Intranets incorporating the feature of online discussion would be a useful knowledge management technology helping communities of practice to share knowledge and promote open flow of information and continual learning.

An intranet is an "internal corporate Internet," or a private network inside a company or organization allowing colleagues to communicate with each other and access corporate information (Chan, 2000). It is also possible to view the intranet as a shared knowledge space for content, communication, and collaboration (Choo, Detlor, & Turnbull, 2000; Stenmark, 2002), as it provides a private space giving the employees in the organization the ability to organize information, readily access that information, manage documents, and enable efficient collaboration, all in a Web-based environment (Intranets.Com, 2004). From a knowledge-sharing point of view, an intranet provides a context where the essence of knowledge-sharing—dialogue, reflection, and perspective-making—could happen (Stenmark, 2002).

Intranets are said to facilitate collaboration, communication, and change mechanisms within organizations, achieving rapid transitions when the pace of change is critical (Ali, 2001). Recent research has shown that an interorganizational virtual team adapted to the use of a collaborative technology (an intranet) and achieved its objective of manufacturing a highly innovative product (Majchrzak, Rice, Malhotra, King, & Ba, 2000). The intranet serves newsgroups that facilitate exchanges of information between members, result-

Table 3. Knowledge management (KM) technologies/practices in smart organizations (SO)

SO Principle	KM Technology/Practice	Features
Value Creating Culture	Extranets	Secure Internet protocol-based networks linking the information infrastructures of various extranet participants.
	Customer Relationship Management Systems	They help manage customer relationships and facilitate relationship marketing.
Creating Alternatives and Disciplined Decision-Making	Intelligent Agents	They can learn the users' behavioural patterns and can have knowledge retrieval, profiling, and filtering capacities.
	Knowledge Interchange Format	It is a language designed for interchanging knowledge among disparate systems.
Alignment, Empowerment, and Systems Thinking	Organizational Decision Support Systems	They are systems providing managers with the relevant internal and external to facilitate their decision-making.
Outside-in Strategic Perspective and Embracing Uncertainty	Enterprise Information Portals	A portal is a single point of access for the pooling, organizing, interacting, and distributing of organizational knowledge.
Open Information Flow and Continual Learning	Communities of Practice	Groups of people are informally bound together by similar expertise for sharing knowledge.
	Intranets	An intranet is an internal corporate Internet for communication and accessing corporate information.

ing in a corporate "knowledge base." The organization's members could subscribe to and view a screen with subject lines, authors, and news articles numbers. Each of these items serves as the beginning of a "thread" that started when someone sent out an article or e-mail; readers can then trace these threads deeper as they wish. In these ways, individuals with the organization can share knowledge and enhance the communication in the community of practice. Employees remain faithful to the informal social networks of "community of practice" as the principal mode of sharing and developing knowledge (Ali, 2001).

Conclusion and Future Direction

As argued above, the common thread running the nine principles characterizing a smart organization is *knowledge*. In order to become smart, an organization needs to have knowledge about the customers' needs and how it can better satisfy their needs in order to create value. Extranets linking the information networks of the organization and the customers could help enhance customer service quality and deliver better value.

Creating alternatives and engaging in disciplined decision-making requires acquiring the knowledge about possibilities and probabilities, evaluating each of them, and selecting the most appropriate option. The proactive intelligent agents facilitate the knowledge discovery and sharing processes, maximizing the knowledge base on which creative ideas are formed and informed decisions are made. Alignment of goals and empowering individuals to act on a shared sense of purpose and deploy systems thinking would be achieved if the individuals within the organization managed complex issues by looking at the cause-and-effect relationships and examining the implications in relation to the big picture. Decision support systems would be able to help by weighing different factors, taking into account the goal and making the optimum choice.

Taking an outside-in strategic perspective and embracing uncertainty would involve gathering knowledge about such environmental factors as the suppliers, competitors, industry, and customers, and utilizing the same to make an appropriate strategic decision. Enterprise portals acting as a single gateway to organizational knowledge sources, both internal and external, would help with collecting information about the uncertain business environment to enable taking an outside-in strategic perspective. Open information flow and continual learning are essential if an organization wants to gain a competitive advantage in the knowledge age, because "knowledge" is the basis of competition; the technology of intranets would facilitate the process.

While there has been a substantial body of literature on knowledge management, as summarized above, recent literature search shows that little research has been done on knowledge management in the context of various aspects of smart organizations such as operations, strategy, human resources, and marketing. What has been proposed above—how knowledge management practices and technologies could help realize each of the nine principles characterizing smart organizations—is based on a theoretical perspective and personal insights rather than empirical evidence. It is based on the author's understanding of and insights into the theories relating to both knowledge management and smart organizations.

This could be regarded as the beginning of setting a future research direction—conducting empirical studies on proving, disproving, extending, or criticizing these theoretical insights to fill in the knowledge gap relating to how knowledge management could make a worthy contribution to smart organizations. For instance, one would look at how the related knowledge management technologies impact recruitment procedures in organizations from a human resources perspective. Another possibility would be examining how knowledge management technologies could innovate supply chain processes. Such studies would make a worthy contribution to the practitioners so that the latter could have a better understanding of how knowledge management technologies generate business value. They would also contribute to academia because little has been done on synergizing knowledge management theories with theories of other management or business disciplines, such as human resources, marketing, or finance. As we are in an era where boundaries are breaking down in various ways—politically, economically, and geographically, for instance—an interdisciplinary approach crossing academic disciplinary boundaries to understand knowledge management and other business disciplines would be timely.

References

Ali, Y. (2001). The intranet and the management of making and using skills. *Journal of Knowledge Management, 5*(4), 338-348.

Ambite, J. L., & Knoblock, C. A. (1997). Agents for information gathering. *IEEE Expert,* September/October, 2-4.

Aneja, A., Brooksby, B., & Rowan, C. (2000). Corporate portal framework for transforming content chaos on intranets. *Intel Technology Journal, 11,* 21-28.

Anonymous (1997). Intellectual capital: Realising your company's true value by finding its hidden brainpower. *Research Technology Management, 40*(5), 59-60.

Band, W. (2000). Creating value for your customer. *Sales and Marketing Management in Canada, 31*(2), 4-6.

Berry, L. L. (1983). Relationship marketing. In G. L. S. Berry & G. D. Upah (Eds.), *Emerging perspectives on services marketing* (pp. 25-28). Chicago: American Marketing Association.

Burns, A., Rathwell, M., & Thomas, R. (1987). A distributed decision-making system. *Decision Support Systems, 3,* 121-131.

Bushko, D., & Raynor, M. (2001). Extranets and client relationships. *Consulting to Management, 12*(2), 3-4.

Caruso, T. E. (1992). Kotler: Future marketers will focus on customer data base to compete globally. *Marketing News, 26*(12), 21-22.

Cascante, L. P., Plaisent, M., Bernard, P., & Maguiraga, L. (2002). The impact of expert decision support systems on the performance of new employee. *Information Resources Management Journal, 15*(4), 64-79.

Chan, S. (2000, September 22). Legal issues for extranets and e-commerce. *Web Review.* Retrieved March 10, 2004, from http://www.catdancers.com/webmags/webrevu/2000/09_22/strategists/09_22_00_2.html

Choo, C. W., Detlor, B., & Turnbull, D. (2000). *Web work: Information seeking and knowledge work on the World Wide Web.* Dordrecht: Kluwer Academic Publishers.

Clarke, I. (2001). Emerging value propositions for m-commerce. *Journal of Business Strategies, 18*(2), 133-149.

Cyert, R. M., & March, J. G. (1963). *A behavioural theory of the firm.* Englewood Cliffs, NJ: Prentice-Hall.

Davenport, T., & Prusak, L. (1997). *Working knowledge: How organisations manage what they know.* Boston: Harvard Business School Press.

Davenport, T. H., De Long, D. W., & Beers, M. C. (1998). Successful knowledge management projects. *Sloan Management Review, 31*(4), 11-27.

Despres, C., & Chauvel, D. (2000). *Knowledge horizons: The present and the promise of knowledge management*. Woburn: Butterworth-Heinemann.

Dess, G. G., & Beard, D. W. (1984). Dimensions of organisational task environments. *Administrative Science Quarterly, 29*(1), 52-73.

Dodgson, M. (1993). *Technological collaboration in industry*. London: Routledge.

Drucker, P. F. (1993). *The post capitalist society*. Oxford, UK: Butterworth-Heinemann.

Duncan, R. B. (1972). Characteristics of organisational environments and perceivedenvironmental uncertainty. *Administrative Science Quarterly, 17,* 313-327.

Earl, M. (1994). The new and old of business process redesign. *Journal of Strategic Information Systems, 3*(1), 5-22.

Earl, M. (2001). Knowledge management strategies: Toward a taxonomy. *Journal of Management Information Systems, 18*(1), 215-233.

Edvinsson, L., & Malone, M. S. (1997). *Intellectual capital: realising your company's true value by finding its hidden brainpower*. New York: Harper Business.

Elenkov, D. S. (1997). Strategic uncertainty and environmental scanning: The case for institutional influences on scanning behaviour. *Strategic Management Journal, 18*(4), 287-303.

Feldman, S., & Yu, E. (1999). Intelligent agents: A primer. *Searcher, 7*(9). Retrieved January 18, 2003, from http://www.infotoday.com/searcher/oct99/feldman+yu.htm

Filos, E., & Banahan, E. (2000). Will the organisation disappear? The challenges of the new economy and future perspectives. In L. M. Camarinha-Matos, H. Afsarmanesh, & R. J. Rabelo (Eds.), *E-business and virtual enterprises* (pp. 3-20). Dordrecht: Kluwer.

Genesereth, M. R. (2004). *Knowledge interchange format*. Retrieved March 10, 2004, from http://logic.stanford.edu/kif/dpans.html

Graham, A. (1996). *The learning organisation: Managing knowledge for business success*. New York: The Economist Intelligence Unit.

Gronroos, C. (1997). Value-driven relational marketing: From products to resources and competencies. *Journal of Marketing Management, 13,* 407-419.

Hayes-Roth, F., Waterman, D. A., & Lenat, D. B. (1983). An overview of expert systems. In F. Hayes-Roth, D. A. Waterman, & D. B. Lenat (Eds.), *Building expert systems* (pp. 3-29). Reading: Addison-Wesley.

Heskett, J. L., Jones, T. O., Loveman, G. W., Sasser, W. E., & Schlesinger, L. A. (1994). Putting the service-profit chain to work. *Harvard Business Review,* March-April, 164-175.

Holsapple, C. W., & Whinston, A. B. (1996). *Decision support systems: A knowledge-based approach.* St. Paul, MN: West Publishing.

Intranets.com (2004). *What is an Intranet?* Retrieved March 10, 2004, from http://www.intranets.com/ProductInfo/FAQs/What_Is_An_Intranet.asp?l=n

Iverson, J. O., & McPhee, R. D. (2002). Knowledge management in communities of practice. *Management communication quarterly, 16*(2), 259-267.

Kendler, P. B. (2000). Portals customise information access. *Insurance and technology, 25,* 47-51.

Kim, G. J., Graves, E. G., Burns, J. R., & Myung, R. (2000). An architecture for organisational decision support systems that utilise a model coordination subsystem. *Journal of End User Computing, 12*(3), 23-35.

Kim, Y. J., Chaudhury, A., & Rao, R. (2002). A knowledge management perspective to evaluation of enterprise information portals. *Knowledge and Process Management, 9*(2), 57-71.

Knoblock, C. A., Arens, Y., & Hsu, C. (1994). Cooperating agents for information retrieval. *Proceedings of the Second International Conference on Cooperative Information Systems.* Toronto, Canada: University of Toronto Press. Retrieved January 18, 2003, from http://www.isi.edu/info-agents/papers/knoblock94-coopis.pdf

Koberg, C. S. (1987). Resource scarcity, environmental uncertainty and adaptive organisational behaviour. *Academy of Management Journal, 30*(4), 798-807.

Kotler, P. (1990). Speech presented at the Trustees Meeting of the Marketing Science Institute, Boston.

Kuhn, T. S. (1970). *The structure of scientific revolution.* Chicago: University of Chicago Press.

Kutner, S., & Cripps, J. (1997). Managing the customer portfolio of healthcare enterprises. *The Healthcare Forum Journal, 4*(5), 52-54.

Lave, J., & Wenger, E. (1990). *Situated learning: Legitimate peripheral participation.* Cambridge, UK: Cambridge University Press.

Lawler, E. E. (1988). Choosing and involvement strategy. *Academy of Management Executive, 2,* 197-204.

MacGregor, R. (1990). The evolving technology of classification-based knowledge representation systems. In J. Sowa (Ed.), *Principles of semantic networks: Explorations in the representation of knowledge.* San Mateo: Morgan Kaufmann.

Majchrzak, A., Rice, R. E., Malhotra, A., King, N., & Ba, S. (2000). Technology adoption: The case of a computer-supported interorganisational virtual team. *MIS Quarterly, 24*(4), 569-584.

Malhotra, Y. (1999). *Who started writing about knowledge management?* Retrieved March 6, 2004, from http://www.brint.com/wwwboard/messages/3332.html

Matheson, D., & Matheson, J. E. (1998). *The smart organisation: Creating value through strategic R&D.* Cambridge, MA: Harvard University Press.

Matheson, D., & Matheson, J. E. (2001). Smart organisations perform better. *Research Technology Management, 44*(4), 49-55.

Miliken, F. J. (1987). Three types of uncertainty about the environment: State, effect and response uncertainty. *Academy of Management Review, 12,* 133-143.

Miller, F. J. (1999). *I = 0. Information has no intrinsic meaning.* Retrieved October 17, 2004, from http://www.sveiby.com/articles/Miller99.html

Nonaka, I., & Takeuchi, H. (1995). *The knowledge-creating company: How Japanese companies create the dynamics of innovation.* New York: Oxford University Press.

O'Keeffe, T. (2002). Organisational learning: A new perspective. *Journal of European Industrial Training, 26*(2-4), 130-142.

Parnell, J. A., Lester, D. L., & Menefee, M. L. (2000). Strategy as a response to organisational uncertainty: An alternative perspective on the strategy-performance relationship. *Management Decision, 38*(8), 520.

Patelis, A., Metaxiotis, K., Nikolpoulos, K., & Assimakopoulos, V. (2003). FORTV: Decision support system for forecasting television viewership. *The Journal of Computer Information Systems, 43*(4), 100-109.

Polanyi, M. (1958). *Personal knowledge*. Chicago: University of Chicago Press.

Polanyi, M. (1966). *The tacit dimension*. London: Routledge & Kegan Paul.

Porter, M. E. (1998). *Competitive advantage: Techniques for analysing industries and competitors*. New York: The Free Press.

Ramirez, R. (1999). Value co-production: Intellectual origins and implications for practice and research. *Strategic Management Journal, 20,* 49.

Raol, J. M., Koong, K. S., Liu, L. C., & Yu, C. S. (2002). An identification and classification of enterprise portal functions and features. *Industrial Management and Data Systems, 102*(7), 390-400.

Rastogi, P. N. (2003). The nature and role of IC: Rethinking the process of value creation and sustained enterprise growth. *Journal of Intellectual Capital, 4*(2), 227-248.

Reichheld, E. (1996). *The loyalty effect*. Boston: Harvard Business School Press.

Rhodes, B. J., & Maes, P. (2000). Just-in-time information retrieval agents. *IBM Systems Journal, 39*(3/4), 685-704.

Roberts, H. (2000). Classification of intellectual capital. In J. E. Grojer & H. Stolowy (Eds.), *Classification of intangibles* (pp. 197-205). Jouy-en-Josas, France: Groupe HEC.

Russell, D., Calvey, D., & Banks, M. (2003). Creating new learning communities: Towards effective e-learning production. *Journal of Workplace Learning, 15*(1), 34-45.

Ryals, L., & Payne, A. (2001). Customer relationship management in financial services: Towards information-enabled relationship marketing. *Journal of Strategic Marketing, 9,* 3-27.

Schneider, A. C., & De Meyer, A. (1991). Responding and interpreting to strategic issues: The impact of national culture. *Strategic Management Journal, 12*(4), 307-320.

Schoderbek, P. P., Schoderbek, C. G., & Kefalas, A. G. (1985). *Management systems: Conceptual considerations*. Plano, TX: Business Publications.

Schroeder, J. (2000). Enterprise portals: A new business intelligence paradigm. *MIS Quarterly, 23,* 124-126.

Senge, P. (1990). *The fifth discipline*. New York: Double Day/Currency.

Sharfman, M. P., & Dean, J. W. (1991). Conceptualising and measuring organisational environment: A multidimensional approach. *Journal of Management, 17,* 681-700.

Sheth, J. N. (2000). Relationship marketing Paradigm shift or shaft? In J. N. Sheth & Parvatiyar (Eds.), *Handbook of relationship marketing* (pp. 609-620). Thousand Oaks: Sage Publications.

Silberston, A. (1967). The patent system. *Lloyds Bank Review, 84*(April 1967), 32-44.

Simons, R. (1995). *Levers of control: How managers use innovative control systems to drive strategic renewal*. Boston: Harvard Business School Press.

Sinkula, J. M. (1994). Marketing information processing and organisational learning. *Journal of Marketing, 58*(January), 35-45.

Skoog, M. (2003). Visualising value creation through the management control of intangibles. *Journal of Intellectual Capital, 4*(4), 487-504.

Snowden, D. (2002). *Complex acts of knowing—Paradox and descriptive self-awareness.* IBM Global Services: July 2003. Retrieved March 10, 2004, from http://www-1.ibm.com/services/files/complex.pdf

Stenmark, D. (2002). Information vs. knowledge: The role of intranets in knowledge management. *Proceedings of the 35th Hawaii International Conference on System Sciences 2002*. Retrieved March 10, 2004, from http://w3.informatik.gu.se/~dixi/publ/ddoml02.pdf

Stenmark, D. (2003). *Research: Agents.* Dick Stenmark's (Volvo Information Technology Department and Viktoria Institute's Knowledge Management Group) Home Page. Retrieved January 16, 2003, from http://w3.informatik.gu.se/~dixi/agent/class.htm

Sveiby, K. (2001). *What is knowledge management?* Retrieved March 1, 2004, from http://www.sveiby.com/articles/KnowledgeManagement.html

Swanston, E. B., & Zmud, R. W. (1989). *Organisational decision support systems: Conceptual notions and architectural guidelines. Information systems and decision processes workshop*. Tucson, AZ.

Ursin, J. (2000). *Group dynamics in the production of new knowledge—A theoretical framework.* Retrieved March 7, 2004, from http://www.leeds.ac.uk/educol/documents/00001609.htm

Vallee, M. (2003). System thinking for quality management and continuous improvement in mining. *CIM Bulletin, 96*(1068), 48-58.

Wenger, E. (1998). *Communities of practice: Learning, meaning and identity*. New York: Cambridge University Press.

Wenger, E. C., & Snyder, W. M. (2000). Communities of practice: The organisational frontier. *Harvard Business Review, 1,* January.

White, M. (2000). Corporate portal: Realising their promises, avoiding costly failure. *Business Information Review, 17,* 71-81.

Yen, D. C., & Chou, D. C. (1999/2000). Extranet: Current developments and future analysis. *The Journal of Computer Information Systems, 40*(2), 46-54.

Chapter V

Bridging Diversity across Time and Space: The Case of Multidisciplinary Virtual Teams

Violina Ratcheva, The University of Sheffield, UK

Abstract

The uniqueness of multidisciplinary teamwork is in its potential to integrate different bodies of knowledge into a new synergy. However, previous empirical studies have shown that member heterogeneity and geographic separation hinder effective sharing and use of team knowledge. The chapter explores how such teams interact to overcome the barriers and take advantage of their "built in" knowledge diversity. The findings indicate that often teams lack common background knowledge at the beginning of the projects, and in order to resolve differences members rely on their external intellectual and social communities. The reported research establishes a positive correlation between team members' participation in multiple professional and social networks and teams' abilities to successfully build on their knowledge diversity. The findings also suggest a need to reconceptualize the boundaries of multidisciplinary teams and to consider the processes of sharing diverse knowledge in a wider social context.

Introduction

With the intensification of globalization and expansion in the use of information technology, particular attention is being focused on the opportunities and difficulties associated with sharing knowledge. The exponential growth of knowledge has made it nearly impossible for any organization to exist in isolation. Thus, the networked organization or alliance is becoming an increasingly common structural form (Leonard, Brands, Edmondson, & Fenwick, 1998). Such networked organizations are usually described as collections of organizations and individuals that have entered into collaborative relations, usually involving multiple channels of communication and knowledge diffusion across disciplinary and organizational boundaries. Previous studies variously refer to such new organizational arrangements as "virtual organizations," "spider's web," "holonic enterprise," "smart organizations," and so forth. Although all describe new ways of organizing that enable people and teams to work across conventional boundaries, there are apparent variations in key characteristics.

A defining component of the virtual organizations, for example, is that they are information computer technology (ICT) enabled (Mowshowitz, 1994) and based on computer-mediated communication (CMC) (Jarvenpaa & Leidner, 1999). Therefore, CMC is a powerful tool to overcome time and distance barriers. It has been recently argued, however, that virtual organizational forms emphasize only one element of what is required from organizations in the digital economy (Filos & Banahan, 2000). To be able to respond to the challenges of the new global marketplace, the organizations have to be not only technologically enabled, but more importantly "smart" in their abilities to enter into virtual collaborations with other partner organizations and share diverse occupational and cultural knowledge. Such "smart organizations" have been described as "organizations that are knowledge-driven, internetworked, dynamically adaptive to new organizational forms and practices, learning as well as agile in their ability to create and exploit the opportunities offered by the new economy" (Filos & Banahan, 2000). Therefore, the knowledge pool and ability to learn, adapt, and be creative are considered as critical for such smart organizations. The main building blocks of such organizations are the multidisciplinary teams working from different locations and team members belonging to different organizations.

The focus on knowledge is particularly acute in the context of the geographically distributed multidisciplinary teams, where the development and delivery of timely and innovative products across heterogeneous cultures and markets are critical and ongoing challenges (Orlikowski, 2002). While the potential advantages of multidisciplinary teams, in terms of creative potential and effectiveness, are theoretically attainable, empirical evidences suggest that knowledge diversity constrains effective sharing (Boutellier, Gassmann, Macho, & Roux, 1998; Gorton & Motwani, 1996; Madhavan & Grover, 1998; Prokesch, 1997; Brown & Eisenhardt, 1995). These constraints have both occupational and contextual origins. Differences in perspectives, priorities, typical approach to problem solving, and professional language can hinder understanding and team cohesion (Doughety, 1992). These difficulties of managing knowledge exchanges amongst team members can become a major barrier to any successful multidisciplinary operation. Previous studies (Doughety, 1992) established that all these "interpretive barriers" might be resolved through team members engaging in highly interactive exchanges.

While previous studies on knowledge processes have examined a variety of settings, most have focused on the work practices of individuals (Orr, 1996) or those of focal groups proximate in time and space (Pentland 1995; Cook & Yanow, 1996). The literature on multidisciplinary teams is very limited and fragmented at this point. Providing a rational-structural definition for this type of teams, previous studies focused mainly on the teams' composition aspects (Nonaka & Takeuchi, 1995; Duarte & Snyder, 1999). For example, Nonaka and Takeuchi (1995) do not attend to the question of how and why knowledge conversions take place and what processes will enhance or interfere with the task's performance. Spender (1998) argues that you cannot talk about knowing (and thus knowledge conversations) without probing the concept of the knower. Looking, therefore, only at the composition of the team is a very limited approach toward understanding the dynamics of multidisciplinary teams where knowledge conversations are taking place.

Little is known about the process of knowing in complex organizations that are also geographically distributed. The complexity, multiplicity, and dispersion of such settings complicate how we think about and study knowledge processes. An important contradiction emerges between the embedded nature of knowledge and the mobility of knowledge in geographically dispersed settings (Clarke & Fujimura, 1992). On one hand, the authentic knowledge processes are somehow embedded within specific practices and interpersonal exchanges; on the other, the successful use of electronic infrastructure to support knowledge processes depends on knowledge being made mobile and transferable across people located in different places.

The paper explores how geographically distributed multidisciplinary teams interact to overcome the communication and cultural barriers and take advantage of their "built-in" knowledge diversity. Dealing with such challenges requires more than just balanced team composition of experts in different fields—it also requires a deep competence in distributed organizing. The focus of the paper, therefore, is on the processes of organizational knowing as emerging from the ongoing and situated actions of team members as they engage with the world, rather than "knowledge" as an outcome of team's activities. In order to be able to address the contradiction between the embedded nature of the processes of collective knowing and the requirement for higher mobility in distributed settings, the author adopts the view that understanding the intra-teams' dynamics requires considering teams in a wider context and acknowledging relationships with various external stakeholders.

The empirical data for this study was gathered through multi-method field research of five dispersed multidisciplinary teams. The findings indicate that often teams lack common background knowledge at the beginning of the projects, and members are accustomed to different working practices. Therefore, in order to resolve differences, members rely for support on their external intellectual and social communities. The reported research establishes a positive correlation between team members' participation in multiple professional and social networks, and teams' abilities to successfully build on their knowledge diversity. The findings also establish a need to reconceptualize the boundaries of multidisciplinary teams and to consider the processes of sharing diverse knowledge in a wider social context.

Factors Affecting Multidisciplinary Operations

Impact of CMC on Information Sharing

Multidisciplinary teams are believed to be useful in developing innovative and optimal solutions to many types of business problems. It has been previously argued that the complementary expertise in multidisciplinary teams can contribute to faster problem solving, enhanced capability to address complex problems in a creative way, and most importantly to create new knowledge about products and processes (Madhavan & Grover, 1998; Boutellier et al., 1998). However, previous studies (Alavi & Yoo, 1997; Cramton, 2001; Cramton & Webber, 1999; Leonard, Brands, Edmondson, & Fenwick, 1998) provide limited insight into the effects of team geographic dispersion. As much of this research stream focuses on the role of technology in supporting remote communication, team members are often selected on the basis of physical location rather than specialized expertise. Previous studies conclude that even in the absence of occupational or functional diversity geographic dispersion can aggravate the complexity of collective work and negatively affect groups' communication (Cramton, 2001).

The geographic dispersion of team members implies heavy reliance on CMC, which can overcome time and distance barriers but suffers from the limitation that nonverbal communication, an important component in trust building, is difficult to achieve. Thus, while nonverbal cues are included in CMC, they are clearly not as easily transmitted as they are in face-to-face communication (Jarvenpaa & Tiller, 1999), and the interpretation of these cues is subject to cultural differences. Previous research has shown that text-based CMC increases the sense of social distance between participants, reduces pressure to conform, and may encourage uninhibited behavior (Bordia, 1997). Therefore, CMC may delay trust formation by slowing the rate at which individuals can gather nonverbal cues about partners' trustworthiness (Bos, Olson, Gergle, Olson, & Wright, 2002). A number of empirical studies comparing information exchange in groups using synchronous text-based computer conferencing and face-to-face groups also support such views (Hightower & Sayeed, 1996; Hollingshead, 1996). They found information exchange to be less complete and discussions more biased in the groups using technology to communicate. One of the most robust findings concerning the effect of computer mediation on communication is that it proceeds at a slower rate than does face-to-face (Straus, 1997). The slower rate has been attributed to the effort required to convey nuances in text without paraverbal and nonverbal cues such as tone of voice, facial expression, and gesture (Hightower & Sayeed, 1996). Therefore, teams communicating through such measures are not able to transmit as much information from their information pool during a given period of time as can groups working face-to-face. As a consequence, less uniquely held information is aired, and their discussion is more biased by commonly held information.

Influence of Diverse Knowledge on Teams' Communications

Information sharing amongst team members is also influenced by differences in working culture, local conditions, work organization practices, access to information, available equipment and support, strength of competing demands, and so forth. For dispersed team members to understand each other and coordinate their work, they must achieve mutual knowledge concerning such contextual differences. Cramton (2001), for example, found that dispersed student teams (connected through technology) with similar educational backgrounds lacked "mutual knowledge" of each member's local context and constraints, and this hindered their ability to work together effectively. By assuming that remote partners experienced the same circumstances they themselves experienced, team members failed to recognize the root causes of miscommunication, and therefore attributed their remote partners' behavior to dispositional rather than situational factors.

The results of previous studies are divided about the impact of geographic dispersion on teams' interactions and performance. Some authors support the view that the demographic diversity is not beneficial in itself, but only to the extent that it represents other diversity, such as of information, values, or perspective (Jehn, Nortcraft, & Neale, 1999). Earlier studies found that demographically diverse groups outperform homogeneous groups (Hoffman, 1978), while too much similarity and group cohesion can result in groupthink, with its associated performance losses. In contrast, some research (Ancona & Caldwell, 1992; Williams & O'Reily, 1998) concludes that increased diversity in working groups and project teams may have dysfunctional effects on group process and performance. In addition, members' similarity in demographic characteristics has been positively associated with team effectiveness and interpersonal attraction (Hambrick & Mason, 1984; Tsui, Egan, & O'Reilly, 1992), and homogeneous members report stronger affinity for their teams than heterogeneous team members (Ibarra, 1992). In general, the relationship between group heterogeneity and performance is mixed, and most likely depends upon task and contextual factors (McGrath, 1984; Williams & O'Reily, 1998).

The lack of agreement about the factors affecting multidisciplinary operations raises a question about how distinctively different such computer-mediated, multidisciplinary interactions are from other forms of network relationships (Staples, Hulland, & Higgins, 1999; Ratcheva & Vyakarnam, 2001; Kraut, Steinfield, Chan, Butler, & Hoag, 1999). The author adopts the view that the geographically dispersed multidisciplinary teams are not simply an evolutionary development of collocated entrepreneurial or new product development teams, and they represent new patterns of interactions. The differences, however, do not purely result from the different locations and variety of communication media used, but more importantly from the different patterns of social exchange, conveying social messages, developing interpersonal and trustworthy relationships, therefore factors which can critically affect the individual willingness to actively share personal knowledge.

Integrating Heterogeneous Knowledge in Multidisciplinary Context

In spite of the apparent advantages of designing teams for knowledge diversity, it is by no means clear how team members make effective use of this knowledge. Grant's (1996) observation, which this paper aims to extend, is that knowledge integration, not knowledge itself, is what generates an advantage for organizations and respectively teams. Penrose (1959) cautioned that the search for knowledge is so voluntary and deliberate, yet so much a part of normal operations, that it cannot be left outside of our system of explanation. Although the organizational form and structure provide the "bones," it is group-level knowledge integration that provides the "flesh and blood" (Van den Bosch, Volberdsa, & Boer, 1999). As new product features are added, new types of specialized knowledge may be required (Penrose, 1959). As new knowledge is brought in on an as-needed basis, it must be integrated with the existing base of knowledge held by the team members. This is perhaps the most compelling explanation for why some teams comprised from the "smartest" and "brightest" experts still fail to perform well. Although the aggregate level of knowledge in such teams might be high, their lack of ability to integrate that knowledge can keep them from gaining any benefits from that resource pool.

Knowledge integration is defined as the project team's ability to continually bring its members' and new external knowledge to collectively bear on the project's execution. Individually held specialist knowledge is synthesized into a new project-specific architectural knowledge. Grant (1996) describes this act as integration, Kogut and Zander (1992), as combination.

Previous research studies indicate that a team's ability to integrate diverse knowledge domains is primarily influenced by the differences and commonalities in the individually held occupational and contextual knowledge. For example, individuals trained in a particular discipline, function, or occupation have substantial conceptual and practical knowledge in common with others from that discipline or occupation (Fleck, 1997). They share terminology and mental frameworks (Vicenti, 1990) which facilitate the efficiency of communication. Different occupations, therefore, have different funds of knowledge (what members know) and systems of meaning (how members know). Dougherty (1992) applied these concepts to organizational departments undertaking product development, and noted that, even though different functional communities were exposed to the same product development circumstances, team members from different functions understood those circumstances differently, selectively perceiving certain aspects as salient and drawing different implications.

Other organizational researchers have focused on the relationships between particular contexts and the unique knowledge acquired in this way (Fleck, 1997; Tyre & Hippel, 1997). Contextual knowledge, therefore, pertains to the broader milieu of the working environment (Fleck, 1997). Many authors have noted the existence of knowledge that resides in systemic routines or ways of interacting, describing such knowledge variously as "organizing principles" (Kogut & Zander, 1992), "embedded knowledge" (Badaracco, 1991; Granovetter, 1985) and "organizing routines" (Levitt & March, 1988; Nelson & Winter, 1982). Contextual knowledge is developed through repetitive collective actions and is "expressed in regularities by which members cooperate in a social community" (Kogut & Zander, 1992). It comprises knowledge of appropriate methods and resources, contributing to communication and task efficiencies and task effectiveness by leveraging taken-for-granted meaning associated with particular behaviour within a specific setting. These associations and behaviors are learned over time from working in a specific setting, and so they are unlikely to be common knowledge among people who are not colocated. In addition, because contextual knowledge tends to be taken for granted by members of a community, it is not easily articulated to members of other communities.

Although previous research acknowledges that different prior knowledge is an integral part of any multidisciplinary operations and team's abilities to integrate, it is a dynamic process, and still little is known about how interactive relationships between team members evolve, develop, and change, and how factors associated with geographic, cultural, and occupational diversity impact such processes.

Theoretical Story Line

The ability of a project team to integrate its members' component knowledge into architectural knowledge influences its ability to execute a project successfully. Given that a larger proportion of component knowledge is held tacitly at an individual level in the form of know-how, specialized skills, and individual expertise, the ability of the multidisciplinary team to integrate it largely determines the extent to which it can bring that knowledge to bear collectively on the project execution. As previously mentioned, the knowledge creation processes are socially constructed, and therefore the articulation of the tacitly held individual knowledge into higher level collectively developed concepts requires an appropriate context which can enable such processes to take place. However, recent studies predominantly focus on enabling context, which resides inside companies' organizational boundaries, and therefore the new knowledge creation processes are well-embedded in the organizational culture, routines, established procedures, and so forth (Nonaka & Konno, 1998). The social interactions in a distributed environment are rather different, and more recently writers started to advocate considering virtualization as a major social process (Diemers, 2000). The virtualization, therefore, requires major reconceptualization of organizational roles, norms, and culture, which traditionally used to constitute the environment in which social interactions took place. In contrast to the "real" environment in which face to face social interactions take place, the virtual networks are only media platform, where according to Harisim (1993) common interpretative spaces of social networks constitute social spaces.

Furthermore, because of the temporal nature of teams, the embedded, tacit practices and routines of working together must be recreated every time because of the nonpersistent social structure. The structural context must compensate for the loss of those social threads (Jarvenpaa & Leidner, 1999). Distributed teams, therefore, require design structure and relationships that are as agile as their markets are dynamic. Unlike collocated teams operating in stable organizational environments that largely depend on learning-before-doing (knowledge stocks), virtually operating teams must also integrate new and emergent knowledge in real time (learning-while-doing). Quick adaptation to market, technological, and environment changes is therefore vital.

Based on the argument that knowledge can be integrated only by teams or groups, the ability of a multidisciplinary team to execute projects successfully will be positively associated with the team's ability to integrate relevant knowledge, expertise, and skills that might be distributed amongst team's members. An underlying assumption in the development of this study was that dispersed multidisciplinary teams represent novel patterns of information exchanges and relationships. The argument, therefore, developed in this study is that multidisciplinary team development and interpersonal processes are likely to follow specific development and adjustment patterns because as members are part of different organizational cultures, they will bring different expectations about work relationships and perceptions of success. Successful knowledge integration processes will be therefore largely determined by three complementary team attributes that together constitute teams' interaction context (see Figure 1): (1) interpersonal interactions and relational capital developed amongst members, (2) work organizational practices and procedures and teams' ability to recognize, interpret, and value information from across its web of participant business units and the external environment, (3) cross-cultural communication behaviours.

Interpersonal Interactions and Developing Relational Capital

Relational capital is defined as level of trust, reciprocity, and strength of ties among the members of a project team (Kale, Singh, & Perlmutter, 2000). Furthermore, relational capital at the team level is influenced by the business units (or organizations) from which those members are drawn. Each unit brings

Figure 1. Conceptual framework

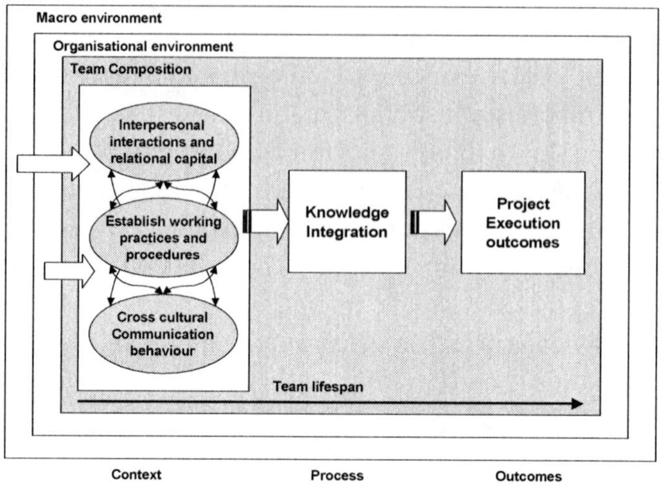

in unique expertise, strengths, and knowledge that must be integrated in the context of the project. As members are drawn from these units to form a project team, they effectively form a web of relationships among individuals that extend to their parent subunits, units, and organizations. This set of linkages can be described as a relational web.

Integrated team capabilities, therefore, depend as much on the individuals' abilities to work together (develop relational capital) as they do on their individual expertise and skills. Strong, trusting, and active relationships within project teams reduce the costs of communication, coordination, and combination of individual expertise at a project level (Kogut & Zander, 1996).

Strong relations facilitate close interactions amongst project team members of different organizational origins (Kale et al., 2000). Close interactions are instrumental in synthesizing sticky, tacit knowledge across organizational boundaries. Higher levels of relational capital are therefore associated with higher levels of learning and knowledge integration between dyadic partners or among members of a relational web.

The willingness of team members consciously and actively to perform their duties critically depends on developing trustworthy relationships. In an environment without formal control and coordination mechanisms, trust has been described as a "heartbeat" which can prevent geographical and organizational distances of team members from turning into unmanageable psychological barriers (Jarvenpaa & Stamps, 1997). The literature acknowledges the existence of impersonal or institutional forms of trust in virtual teams, in addition to interpersonal forms. According to Luhmann (1986), impersonal trust is based on the appearance of "everything in proper order," rather than on an emotional bond, knowledge, or past history of interactions. Meyerson, Weike, and Kramer (1996) developed the concept of "swift" trust to explain how temporary teams can enjoy high levels of trust, even though members do not share any past affiliation and cannot necessarily expect to have any further associations. The concept of "swift" trust maintains that "unless one trusts quickly, one may never trust at all" (Meyerson et al., 1996). Because there is not a sufficient time to develop trust through interpersonal means, team members import expectations of trust based on their local organizational environment, industry practices, or role-based stereotypes. Positive expectations of trust motivate members to take a proactive part in the team, which can result in strengthening the trustworthy relationships amongst team members. Relational capital, therefore, serves as an effective, cost-efficient, and self-enforcing mechanism that improves knowledge integration while simultaneously discouraging opportunism in a relationship web.

Working Practices and Routines

Time, interaction, and performance theory (TIP) (McGrath, 1991) represents an emerging trend in small group research that takes into account the new, temporally oriented tradition. Work in this tradition treats groups dynamically and attempts to take full account of the physical, temporal, and social context within which those groups are embedded.

McGrath's (1991) TIP theory describes work groups as time-based, multi-functional, and with multi-modal social systems. Effective groups are engaged simultaneously and continuously in three functions: (1) production (problem solving and task-performance), (2) member-support (member inclusion, participation, loyalty, and commitment) and (3) group well-being (interaction, member roles, power, and politics). Teams carry out three functions by means of activity that relate to four possible modes: (Mode 1) inception and acceptance of a project, (Mode 2) problem solving, (Mode 3) conflict resolution, and (Mode 4) project execution. The modes and functions, according to McGrath and Hollingshead (1994), are not a fixed sequence of phases, but dependant on the team, tasks, technology, time, and other environmental contingencies. The TIP theory suggests that a team with no past history that is working on a challenging problem with much technological and environmental uncertainty has to engage in all four functions and modes to avoid detrimental effects on performance. Multiple involvement in various functions and tasks and, therefore, low division of roles in a virtual team can enhance the team's integrity and consequently enhance the team's performance.

Influence of Cultural Differences on Communication Behaviours

The culture-specific behaviours of individuals brought to work on cross-organizational projects are also likely to have a significant impact on team's integrity (Gudykunst, 1997). One dimension of a cultural variability identified to have an impact on ability to work successfully in a team is individualism-collectivism (Hofstede, 1991). In individualistic cultures, the needs, values, and goals of the individual take precedence over the needs, values, and goals of the in-group. Opposite, in collectivist cultures, the needs, values, and goals of the in-group take precedence over the needs, values, and goals of the individual (Gudykunst, 1997). Empirical findings suggest also that individuals from individualistic cultures tend to be less concerned with self-categorizing, they are less influenced by group membership, have greater skills in entering and leaving new groups, and engage in a more open and precise communication than individuals from collectivist cultures (Hofstede, 1991). Previous cultural exposure has also been identified as an important factor influencing communication behavior.

Team members import working behavior from other settings with which they are familiar. Therefore, virtual team members representing organizations with strong collectivist cultures are more likely to develop strong social bonds over a longer period of time and less likely to involve in a more open communication in contrast to individuals representing individualist organizational cultures.

Research Methodology and Sample Definition

Five small- and medium-size companies in the UK took part in a longitudinal qualitative study investigating the team development processes which enable or hinder the diverse knowledge integration in geographically dispersed multidisciplinary context. The companies represent technologically advanced industries and provide a wide range of engineering and software services. The results presented in this paper are the preliminary outcomes of the second stage of the research project specifically focusing on successful practices in integrating diverse knowledge which resulted in novel products, procedures, processes, and so forth. A common characteristic of the sample companies is that they went through major strategic and structural change processes during the late 1990s in order to maintain their competitive positions. These change processes revolved around a redefinition of the vision and identification of key areas where innovations and work process improvements could continually support the companies' strategic edge (see Table 1 for companies' background information). One of the outcomes of the restructuring initiatives was the increased reliance on multidisciplinary virtual teams to handle a variety of business initiatives, formed across organizational and country boundaries.

The present study was carried out using a multi-method approach. The companies selected were initially considered as a focal point for identifying project partnerships. Each company was asked to identify a multidisciplinary partnership in which the particular organization had played a leading role in terms of resource commitment, and the outcomes of the partnership were highly satisfactory. In order to maintain consistency between cases, the teams were selected according to the following criteria:

Table 1. Companies' background

Cases	Main activities	Team boundaries	No. of team members
Case 1	Engineering and software project consulting	Different organisations, operating in 2 countries	8
Case 2	Engineering consultancy	Different organisations, operating in 3 countries	7
Case 3	Electronic modem assembly	Different organisations, operating in 2 countries	9
Case 4	Assembly of electronic connectors	Different organisations, operating in 3 countries	10
Case 5	Research and development engineering consultancy	Different organisations, operating in 3 countries	7

- Teams included members with diverse expertise (different functional or subject areas).
- A variety of communication channels were used with electronic communications being the main method throughout the lifespan of the project.
- Teams included members from more than two organizations (or independent experts) working on the project from different geographical locations.
- Teams working on projects, the outcomes of which were considered by the companies as highly successful in terms of new knowledge creation resulting in highly innovative outcomes.

In order to achieve consistency between cases, the collective knowledge created in each partnership was measured using the Innovation Assessment Questionnaire previously used by Sethi (1995). Further evaluation was carried out using a creativity scale (Andrew & Smith, 1996), which measured the originality of the project outcome (novelty dimension) and its usefulness (appropriateness dimension). The partnerships that took part in the study had high scores for both novelty and usefulness.

The identified five multidisciplinary teams were further investigated in depth using a variety of data collection approaches. The data were analysed using content analysis and a coding scheme procedure (Weber, 1985) in order to illuminate the underlying differences between the partnerships and identify the key factors and processes affecting the teams' abilities to create new collective knowledge.

Research Findings

The initially approached five companies showed a trend toward breaking with an old tradition of developing business. A common characteristic is that they currently develop their products or perform their operations relying entirely on partnerships which required implementation of profound changes in the organizational strategy, structure, and everyday routines. Two distinctive paths in their developments were identified.

In three of the cases, the businesses were formed around an exciting idea and realized potential of working jointly across business boundaries. The team members, who were former university colleagues and computer scientists in one of the cases, and in another case were professionals with a long standing experience working for large corporations and who felt that they did not really belong there, came together as teams because of their similar educational and professional backgrounds with clear understandings of how to exploit the potential of information technologies. As one of the respondents stated, "our futures depends on opportunities spotters rather than marketers."

In contrast, the other two businesses went through a long period of organizational and cultural adjustment to find a new way of radical thinking (complete changes of the product/service offered, sharing information and resources, learning to work and trust people they had never met before, etc.), either forced by changing trends in the particular industry or by a new generation of the family taking over the business who was no longer excited with the founders' legacy. In both companies there was an understanding among the management team that the changing directions contradicted with some of the traditional working values. In one of the cases the changes have been accomplished by appointment of a new management team, and in the other by training and development of the key staff over several years.

However, no significant differences were observed related to the factors triggering the initial formation of the investigated teams in which the above companies had key involvement. A common trend is that an opportunity is spotted or idea arises before the team is formed. The teams were formed in order to accomplish a particular project, and therefore the team selection in all of the cases reflected on personal skills and knowledge. A similarity amongst teams was that both occupational and contextual knowledge sources were acknowledged and sought out at the teams' formation stage. Initially, when staffing the project teams, the focus was on occupational knowledge. Consistent with past research in product development, these teams were expected to leverage expertise of diverse functions and scientific fields to accomplish challenging development needs.

Clear occupational belonging of team members proved to be an influential factor during teams' initiation and formation. Because of the temporary nature of the projects, team members import to the partnerships their perception and understanding about each other's potential to contribute in terms of having an appropriate occupational knowledge. For example, "He's electrical engineer" or "I'm an experimental scientist" acted as shorthand for conveying information about distinct skills, expertise, and conceptual insights that someone might bring to bear on a problem. The relationships building at that stage were based on the potential to contribute unique personal knowledge and are highly depersonalised. Team members interviewed stated that "what others can bring to the project rather than how we feel about working with each other" are the most important initial selection criteria. Positive expectations of members' valuable occupational knowledge, therefore, motivated participants to take a proactive part in the team, which resulted in strengthening the trustworthy relationships amongst team members and contributed to establishing more active knowledge sharing practices.

At the same time, a clear tendency to access help particular to a specific site led to considering the locally based or contextual knowledge as a new and distinct source of value for the development teams. In some cases, understanding the context in which the product was targeted for use was critical for making appropriate social and cultural decisions regarding its design and implementation. Whenever possible, potential users of the product were intentionally sought to represent that unique viewpoint within the team. On a number of occasions, contextual knowledge simply entailed knowing who to contact for further advice or resources in order to accomplish certain objectives.

Interesting relationships were observed among team members with clear occupational and others with contextual knowledge. Although acknowledged, that understanding of the context in which the product/consultancy advice was targeted for use was critical for making appropriate social and cultural decisions, the occupational specialists were considered as the "knowledgeable" hard core of the teams and to rest was referred as "social agents" brought in on ad-hoc basis.

Teams which realized early in the formation stage the importance of continuously combining contextual and occupational knowledge throughout the lifespan of the partnership were more flexible in redirecting the project by recombining knowledge according to external, environmental changes such as changing customer requirements, new competitive offerings, new technological advances, and so forth. These external changes led to redefinition of roles and responsibilities in the team and introduction of complementary external expertise as required. This caused further changes in the team's patterns of interaction and knowledge base. Therefore, ensuring appropriate mix of expertise throughout the lifespan of the project contributed to the progress of the projects, which increased members' confidence in the ability of the team.

The results of the study also confirm that social and personal relationships within a given local (physical) or virtual community were extremely effective in gaining team members "just-in-time" access to specialist knowledge and practical skills as and when the team needed it. Through such boundary-spanning activities, project teams gained access to broader and deeper skills and expertise, which helped in addressing specific project issues. Team members, for example, regularly pulled in their collocated colleagues for assistance with practical advice or input on decisions, all of which enabled the projects to proceed to the next stage. Seeking assistance from a wider community, however, was more than just seeking additional feedback and task-related assistance, but was also looking for moral support when faced with difficult decisions. Therefore, the intellectual, virtual, and colocated communities of which members are part became an integral part of the thinking and discovery processes and filled knowledge gaps by contributing timely and efficient access to broad expertise, practical assistance, and emotional support which were not internally available.

The findings also raise a number of questions such as: What is the impact of cross-cultural differences on developing collective knowledge? What novel theoretical approaches can capture the complexity of such relationships? How can we and should we separate in our analyses the team members from the organizations they represent and the wider community in order to gain an in-depth insight? How to transfer the successful experience from one partnership to another?

A limitation of the reported results is that initially the participating companies, rather than the partnerships, were approached. The partnerships were identified based on the companies' preferences, which may have affected the validity of the results. Another limitation of this study is that because of the geographical distance of the partners involved not everybody associated with a particular project was approached during the data collection. In that respect, the results presented here should be treated as preliminary.

References

Alavi, M., & Yoo, Y. (1997). *Is learning in virtual teams real?* Boston: Harvard Business School.

Ancona, D., & Caldwell, D. F. (1992). Demography and design: Predictors of new product team performance. *Organization Science, 3,* 321-341.

Andrew, J., & Smith, D. (1996). *Getting beyond me-too marketing: Determinants of the creativity of marketing programs for mature products.* Unpublished manuscript, Case Western Reserve University.

Badaracco, J. (1991). *The knowledge link*. Boston: Harvard Business School Press.

Bordia, P. (1997). Face-to-face versus computer-mediated communication: A synthesis of experimental literature. *Journal of Business Communications, 34*(1), 99-120.

Bos, N., Olson, J., Gergle, D., Olson, G., & Wright, Z. (2002, April). *Effects of four computer-mediated communications channels on trust development.* Paper presented at CHI, Minneapolis, MN, USA.

Boutellier, R., Gassmann, O., Macho, H., & Roux, M. (1998). Management of dispersed product development teams: The role of information technologies. *R&D Management, 28*(1), 13-26.

Brown, S., & Eisenhardt, K. (1995). Product development: Past research, present findings, and future directions. *Academy of Management Review, 20*(2), 343-378.

Clarke, A. E., & Fujimura, J. H. (1992). *The right tools for the right job: At work in 20th century life scenes*. Princeton, NJ: Princeton University Press.

Cook, S. & Yanow, D. (1996). Culture and organizational learning. In M. D. Cohen & L. S. Sproull (Eds.), *Organizational learning.* Thousand Oaks, CA: Sage Publications.

Cramton, C. D. (2001). The mutual knowledge problem and its consequences for dispersed collaboration. *Organization Science, 12*(3), 346-371

Cramton, C. D., & Webber, S. S. (1999). *A model of the effects of geographical dispersion on work teams*. Fairfax, VA: George Mason University.

Diemers, D. (2000). Information quality and its interpretative reconfiguration as a premise of knowledge management in virtual organisations. In Y. Malhotra (Ed.), *Knowledge management and virtual organizations* (pp. 365-379). Hershey, PA: Idea Group Publishing.

Dougherty, D. (1992). Interpretative barriers to successful product innovation in large firms. *Organization Science, 3*(2), 179-202.

Duarte, D. L., & Snyder, N. T. (1999). *Mastering virtual teams*. San Francisco: Jossey-Bass Publishers.

Filos, E., & Banahan, E. (2000). Will the organisation disappear? The challenges of the new economy and future perspectives. In L. M.

Camarinha-Matos, H. Afsarmanesh, & R. J. Rabelo (Eds.), *E-business & virtual enterprises* (pp. 3-20). Dordrecht: Kluwer.

Fleck, J. (1997). Contingent knowledge and technology development. *Technology Analysis and Strategic Management, 9*(4), 383-398.

Gorton, I., & Motwani, S. (1996). Issues in co-operative software engineering using globally distributed teams. *Information and Software Technology, 38*(10), 647-656.

Granovetter, M. (1985). Economic action and social structure: The problem of embeddedness. *American Journal of Sociology, 91*(3), 481-510.

Grant, R. (1996). Toward a knowledge-based theory of the firm. *Strategic Management Journal, 17*(Winter Special Issue), 109-122.

Gudykunst, W. B. (1997). Cultural variability in communication. *Communication Research, 24*(4), 327-347.

Hambrick, D. C., & Mason, P. A. (1984). Upper echelons: Organisation as a reflection of its top managers. *Academy of Management Review, 9*(2), 192-206.

Harisim, L. M. (1993). Networlds: Networks as social space. In L. M. Harasim (Ed.), *Global networks, computers and international communication* (pp. 15-45). Cambridge, MA: MIT Press.

Hightower, R., & Sayeed, L. (1996). Effects of communication mode and prediscussion information distribution characteristics on information exchange in groups. *Information Systems Research, 7*(4), 451-465.

Hoffman, L. R. (1978). The group problem-solving process. In L. Bercowitz (Ed.), *Group processes* (pp. 101-114). New York: Academic Press.

Hofsteade, G. (1991). *Cultures and organisations: Software of the mind.* McGraw-Hill

Hollingshead, A. (1996). Information suppression and status persistence in group decision making: The effects of communication media. *Human Communication Research, 23*(2), 193-219.

Ibarra, H. (1992). Homophily and differential returns: Sex differences in network structure and access in an advertising firm. *Administrative Science Quarterly, 37*(3), 422-447.

Jarvenpaa, S., & Leidner, D. E. (1999). Communication and trust in global virtual teams. *Organization Science, 10*(6), 791-827.

Jarvenpaa, S., & Tiller, E. (1999). Integrating market, technology, and policy opportunities in e- business strategy. *Strategic Information Systems, 8*(3), 235-249.

Jehn, K. A., Nortcraft, G. B., & Neale, M. A. (1999). Why differences make a difference: A field study of diversity, conflict, and performance in workgroups. *Administrative Science Quarterly, 44*(4), 741-763.

Kale, P., Singh, H., & Perlmutter, H. (2000). Learning and protection of proprietary assets in strategic alliances: Building relational capital. *Strategic Management Journal, 21,* 217-37.

Kogut, B., & Zander, U. (1992). Knowledge of the firm, combinative capabilities, and the replication of technology. *Organization Science, 3*(3), 383-397.

Kraut, R., Steinfield, C., Chan, A. P., Butler, B., & Hoag, A. (1999). Coordination and virtualization: The role of electronic networks and personal relationships. *Organization Science, 10*(6), 722-740.

Leonard, D. A., Brands, P. A., Edmondson, A., & Fenwick, J. (1998). Virtual teams: Using communications technology to manage geographically dispersed development groups. In S. P. Bradley & R. L. Nolan (Eds.), *Sense and respond: Capturing value in the network era* (pp. 285-198). Boston: Harvard Business School Press.

Levitt, B., & March, J. G. (1988). Organizational learning. *Stanford University Annual Review of Sociology, 14,* 319-340.

Lipnack, J., & Stamps, J. (1997). *Virtual teams: Reaching across space, time and organisations with technology*. New York: John Wiley & Sons.

Luhmann, N. (1986). The autopoiesis of social systems. In F. Geyer & J. Van der Zouwen (Eds.), *Sociocybernetic paradoxes* (pp. 172-192). Beverly Hills, CA: Sage.

Madhavan, R., & Grover, R. (1998). From embedded knowledge to embodied knowledge: New product development as knowledge management. *Journal of Marketing, 62*(4), 1-12.

McGrath, J. E. (1984). *Groups: Interaction and performance*. Englewood Cliffs, NJ: Prentice-Hall.

McGrath, J. E. (1991). Time, interaction and performance (TIP): A theory of groups. *Small Group Research, 22*(2), 147-174.

McGrath, J., & Hollingshead, A. B. (1994). *Groups interacting with technology.* Thousand Oaks, CA: Sage Publications.

Meyerson, D., Weick, K. E., & Kramer, R. M. (1996). Swift trust and temporary groups. In R. M. Kramer & T. R. Tayler (Eds.), *Trust in organisations: Frontiers of theory and research* (pp. 166-195). Thousand Oaks, CA: Sage.

Mowshowitz, A. (1994). Virtual organisation: A vision of management in the information age. *Information Society, 10,* 267-288.

Nelson, R., & Winter, S. (1982). *An evolutionary theory of economic change.* Cambridge, MA: The Belknap Press of Harvard Business School.

Nonaka, I., & Konno, N. (1998). The concept of 'ba': Building a foundation for knowledge creation. *California Management Review, 40*(3), 40-54.

Nonaka, I., & Takeuchi, I. (1995). *The knowledge creating company: How Japanese companies create the dynamics of innovation.* New York: Oxford University Press..

Orlikowski, W. J. (2002). Knowing in practice: Enacting a collective capability in distributed organizing. *Organization Science, 13*(3), 249-273.

Orr, J. (1996). *Talking about machines: An ethnography of a modern job.* Ithaca, NY: ILP Press.

Penrose, E. T. (1959). *The theory of the growth of the firm.* New York: Wiley & Sons.

Pentland, B. (1995). Information systems and organizational learning: The social epistemology of organizational knowledge systems. *Accounting, Management and Information Technology, 5,* 1-21.

Prokesch, S. E. (1997). Unleashing the power of learning: An interview with British Petroleum's John Browne. *Harvard Business Review, 75*(5), 147-164.

Ratcheva, V., & Vyakarnam, S. (2001). Building virtual relationships in a distributed organisational environment. In Y. Malhotra (Ed.), *Knowledge management and business model innovation* (pp. 170-201). Hershey, PA: Idea Group Publishing.

Sethi, R. (1995). *New product development and cross-functional teams.* Unpublished doctoral dissertation, University of Pittsburgh.

Spender, J. (1998). *Pluralist epistemology and the knowledge based theory of the firm*. Unpublished manuscript, New York Institute of Technology.

Staples, D. S., Hulland, J. S., & Higgins, C. A. (1999). A self-efficacy theory explanation for the management of remote workers in virtual organizations. *Organization Science, 10*(6), 758-776.

Straus, S. (1997). Technology, group process, and group outcomes; Testing the connections in computer-mediated and face-to-face groups. *Human-Computer Interaction, 12,* 227-266.

Tsui, A. S., Egan, T. D., & O'Reilly, C. A. (1992). Being different: Relational demography and organizational commitment. *Administrative Science Quarterly, 37*(4), 549-579.

Tyre, M. J., & Hippel, E. V. (1997). The situated nature of adaptive learning in organisations. *Organization Science, 8*(1), 71-83.

Van den Bosch, F., Volberdsa, H., & Boer, M. (1999). Convolution of firm absorptive capacity and knowledge environment: Organisational forms and combinative capabilities. *Organisation Science, 10*(5), 551-568.

Vicenti, W. G. (1990). *What engineers know and how they know it: Analytical studies from aeronautical history*. Baltimore; London: John Hopkins University Press.

Weber, R. (1985). *Basic content analysis*. Thousand Oaks, CA: Sage.

Williams, K. Y., & O'Reily, C. A. (1998). Demography and diversity in organisations. In B. M. Staw & R. M. Sutton (Eds.), *Research in organizational behaviour* (pp. 77-140). Stamford, CT: JAI Press.

Chapter VI

Neural Data Mining System for Trust-Based Evaluation in Smart Organizations

T. T. Wong, The Hong Kong Polytechnic University, Hong Kong

Abstract

Nowadays, many enterprises manufacture and distribute their products or services globally, and quite a number of smart organizations are formed on the Internet and are expected to evolve to a strategically important e-business model. Although information and communication technologies (ICT) and knowledge management plays an important role in linking the core and partner companies, it remains subservient to the humans that form the smart organizations. This chapter identifies two instances in which trust-based evaluations of partners in the smart organizations are applicable. A review of the literature indicates that neither researchers nor practitioners agree on a single model of interfirm trust that applies to all partner evaluation contexts. A decision-support system based on neural network and data mining technologies is proposed. A case example is given to illustrate a trust-based evaluation in real situation.

Introduction

Fostered by the rapid and ever-increasing pace of development of information and communications technologies (ICT), a new digital economy is emerging around the globe. In this new digital economy, small and medium enterprises will have more chances than ever before to play a role alongside the big players. Since the Internet is being adopted as the most important global communications infrastructure, and because global knowledge is becoming accessible by everyone with Internet access, the digital economy will be characterized by radically new forms of business relationships and will have a profound effect on the way we work and live.

The current business environment can be typified by continuous pressures to change and the shortening of product development times and product life cycles. Markets are fickle and strongly customer-led, with unrelenting pressures to bring down costs and to take advantage of innovative technologies. Many organizations look to new organizational structures such as the smart organization to enable them to cope with these changed conditions. As computers rapidly shrink in size, ICT are becoming ubiquitous and invisible. Organizations preparing to go digital need to recognize the implications this transition will have on their business processes and organizations as a whole. Besides setting up the digital infrastructure, organizations need to be able to:

- enter into virtual global internetworked collaborations
- manage transition and adapt quickly
- leverage human capital to optimize performance

In other words, organizations need to get "smart." The characteristics of a smart organization (Filos & Banahan, 2000) adopted in the present context are:

- internetworked
- dynamically adaptive to new organizational forms and practices
- knowledge and learning-driven
- hierarchically flattened where the individual's skills, intellect, and knowledge, are recognized, valued, and leveraged

In the past, it was more cost-effective to own all aspects of the value chain—vertical integration was the business model of choice. In today's global market, focus is critical. Owning the value chain may actually put an organization at a competitive disadvantage due to the lack of flexibility and financial commitment true vertical integration represents. Selecting the right partners and nurturing these relationships can help a company focus on what creates the most value for customers and concentrate on its core activities. Smart organizations also offer versatility. They create new, viable market options and allow companies to deal more effectively with the uncertainties and complexities of today's highly competitive global market.

Following Jarvenpaa and Leidner (1998), a global smart organization can be defined as a temporary, culturally diverse, geographically dispersed, electronically communicating work group. The notion of temporary in the definition describes teams on which members may have never worked together before and who may not expect to work together again as a group (Lipnack & Stamps, 1997). The characterization of smart organizations as global implies culturally diverse and globally spanning members that can think and act in concert with the diversity of the global environment (DeSanctis & Poole, 1997). Finally, it is a heavy reliance on the integration of information and communication technologies (ICT), knowledge and organizational networks that allows members separated by time and space to engage in collaborative work. The reasons that smart organizations are becoming so prevalent nowadays include low overhead, flexibility, minimum investment, and high productivity. By owning few resources and focusing on the organization's expertise, the company can keep high levels of productivity while allowing its partners to do the same. Both the partners in a smart organization and the individuals who work for the partners are allotted greater flexibility. The partners can focus on core competencies, while individual workers may have the ability to telecommute from their homes. In a smart organization, companies are linked by the free flow of information. There is no hierarchy, no central office, and no vertical integration: just the skills and resources needed to do the job. Each participating company contributes what it is best at. It can be seen that since no single company will have all the skills necessary to compete in the global electronic market, these arrangements will become the norm. One of the keys to the success of the smart organization is the use of ICT to facilitate these alliances.

Creating a smart organization takes more than just the ICT. A recent study on issues of information technology and knowledge management concluded that there is no evidence that ICT provides options with long-term sustainable competitive advantage. The real benefits of ICT derive from the constructive combination of ICT with organization culture for learning and innovation, supporting the trend toward more flexible forms of organization. Information technology's power is not in how it changes the organization, but in the potential it provides for allowing people to change themselves. Creating these changes, however, presents a whole new set of human issues. Among the biggest of these challenges is the issue of trust in system security and trust among partners in the smart organization. It has been reported that the lack of trustworthy security services is a major obstacle to the use of information systems in private business (B2B) as well as in public services, and trust is closely linked to consumers' rights, such as security, identification, authentication, privacy, and confidentiality (Mezgar & Kincses, 2003).

In the following sections the importance of trust in the operation of a smart organization will be described, and suggestions on trust building and discussion of the trust problems associated with ICT system security and smart organization partnering will follow. A trust-based evaluation of smart organization partners using the neural network and data mining technologies will be given.

Importance of Trust in Smart Organizations

While technology seems to provide sufficient safeguards, such as digital signatures, encryption, web seal assurances, and standards that provide technology-based security and trust mechanisms at present, there seems to be a perception by businesses that transactions conducted specifically via the Internet are insecure and unreliable (Bhimani, 1996; Raman, 1996). The findings of a recent electronic commerce survey that examined the adoption of electronic commerce in Australia and New Zealand indicated that there is still an initial slow growth of Internet business to business electronic commerce within the Asia Pacific region as compared to United States. This is due to perceived risks in the security of business to business electronic commerce transactions and trust of the trading parties (KPMG, 1999). It seems as if despite the technical assurances, trading partners in business to business electronic commerce do not trust the "people side" of the transactions. Parkhe (1998) suggests that this is due to two types of uncertainties: uncertainty regarding unknown future events, and uncertainty regarding trading partners' responses to future events. It is in this environment of dual uncertainty that trust becomes an important element in the effective operation of smart organizations.

Trust has long been of interest to a variety of researchers. Mayer, Davis, and Schoorman (1995) define trust as "the willingness of a party to be vulnerable to the actions of another party based on the expectation that the other will perform a particular action important to the trustor, irrespective of the ability to monitor or control that other party." Trust among partners is critical for smart organizations. Without trust, commitment to the goals of the virtual organization can waver, as members perceive the alliance as weak or disintegrating, fractured by misunderstanding or mistrust (Handy, 1995). Trust is particularly important in networked organizations that require constant and close attention to shared commitments to safety and reliability, as well as a shared willingness to learn and adapt. It has been suggested that trust permits an organization to focus on its mission, unfettered by doubts about other members' roles, responsibilities, and resources, and that with trust, synergistic efforts in interorganizational missions are possible (Gabrowski & Roberts, 1998). Developing trust in a smart organization is a complex task. It requires fairly constant, small group activities among partners, because it is difficult to trust people you do not know well, whom you have not observed in action over time, and who are not committed to the same goals. Trust plays an important synthesis role because with trust, smart organization can leverage the ability and willingness to learn (Coyle & Schnarr, 1995), thereby enhancing performance and attention to reliability over time. Smart organizations with high levels of trust among their members can effectively utilize interactions and communication processes at their interfaces so partners can learn together, and can develop shared mental models of reliability and a shared culture of innovation. High levels of trust also contribute to strengthening linkages among member organizations. Trust among organizational members is an important prerequisite to changing these linkages to alliances, thus mitigating risks, as organizations are reluctant to adopt alliance-like organizational structures that make them vulnerable to the uncertainties of the global market, and to impacts from other organizations, without some assurances of shared vulnerability (Handy, 1995; McAllister, 1995). Aerospace conglomerates jointly developing mission- and safety-critical systems, comprised of defense contractors, universities, government departments, and other private organizations, are good examples of the need for trust in smart organizations. In these organizations, members have different backgrounds, experiences, goals, objectives, and understandings, but belong to the conglomerate to pursue shared development of mission- and safety-critical aerospace systems. For effective smart organization management to take place, however, members needed to trust in the other members' talents, capacities, willingness to work, and interests in the alliance (Augustine, 1997; SmartBridge, 1997). Without such trust, members could duplicate other members' efforts, subvert the mission and goals of the conglomerate by providing private information about conglomerate members to the members' competitors, and introduce inefficiencies and costs pathological to the smart organization goals.

Trust is one of the key drivers for innovation. Many researchers support the point that trust is likely to favor technological innovation and change (Humphrey & Schmitz, 1998; Lorenz, 1999; Moore, 1999; Maskell, 2000).

Trust Building

From the above it is seen that the building and maintenance of trust is a prerequisite for successful smart organization operation. Given the diversity among the smart organization partners it may be unrealistic to seek universal solutions on a one-size-fits-all basis. From the literature the essential conditions for trust development in a smart organization may be summarized as follows:

A Common Business Understanding

Researchers such as Wigand et al. (1997) and Fuehrer and Ashkanasy (2001) noted that a crucial element in any business cooperation is the establishment of a common business understanding. An earlier work suggests that there are three specifications necessary for the establishment of a common business understanding in the networked context. The first is a clear product specification: the design, quality, and functionality of the product or service. The second is specification of the cooperation level, which requires agreement about deadlines, liability, prices, profit allocation, and staff and resource input. The third is formal specification of agreements between the partners. In a smart organization, these specifications need to be communicated clearly among all partners to achieve a common business understanding. There is always varying uncertainty between members, however. Therefore, there is a need to guard against opportunistic behavior varies between the partners (Wicks et al., 1999).

The three specifications (production, cooperation, and agreements between partners) can be achieved by negotiating relational contracts that guide the formation, operation, and dissolution of the smart organization, thereby facilitating an increase in the level of collaboration-enabling trust. Smart organizations, like other networked organizations, create fiscal and legal issues that must be clarified but they lack a formalized legal framework. Therefore, it is incumbent on the organization's members to develop their own guidelines for the operation of the enterprise. Such agreements may include clarification of members' tasks and responsibilities, agreement on contracts, allocation of funds, potential liability, and how members will contribute their expertise. In this sense, clear guidelines, spelled out in an early stage of the partnership, serve to reduce misperceptions and to foster the establishment of trust.

Other mechanisms to establish a common business understanding in a smart organization include development of an organization handbook, design of a mutual Internet site, chat room technology, or the use of team addresses for e-mail. A specific example is Livelink, a software selected by Siemens to enable creation of a common business understanding for virtual collaboration and knowledge-sharing across geographical and organizational boundaries.

High Ethical Standards

Three factors uniquely characterize the smart organization's position in regard to business ethics. Firstly, smart organizations are rarely guided by pre-existing codified laws, where values and standards are written into legal systems enforceable in court. Since the partners are not usually legally bound to the organization, any negative outcomes or perceptions attributed to poor business ethics could result in the organization's reputation suffering (Fichman, 1997). Second, because smart organizations are intrinsically temporary, corporate ethics are difficult to develop because members will typically be finishing one Internet collaboration and entering into another. Thirdly, smart organizations are intrinsically boundary-spanning in nature, so that they must incorporate a diversity of culturally based values and morals.

Researchers focused on the notion of advances in ICT and the related effects on social behavior agree that unethical behavior in the ICT context is predominantly caused by technological changes and by the "inside keepers of the information systems" (Pearson et al., 1997, p. 94). They also agree that social behavior needs more than "new laws and modified edicts" (Johnson, 1997, p. 60), and that ethical issues will become increasingly important to enable business transactions to be carried out safely and securely. Although technology has been largely secured by advancing software and technology for virus detection, as well as en/decryption of information to ensure the security of business processes, Johnson (1997) noted that technology could never be sufficient to control all aspects of social behavior. Consequently, online behavior is predicated on an awareness and acceptance of ethical norms and behaviors. This can best be achieved through specification and clarification of the members' tasks, responsibilities, and agreed sanctions for proscribed behavior.

Pearson et al. (1997) reported on ethical standards for the IS profession proposed by three major professional associations in this field. These associations share an agreed set of behavioral obligations to society, to colleagues, and to professional organizations. The standards aim to promote the principle that individuals within the professions act in an ethical and responsible manner in order to influence the success of their organizations (Pearson et al., 1997). Clearly, similar standards can be developed for the operation of smart organizations specifying, for instance, the obligation to partners and clients.

Other possible mechanisms to promote ethical behavior in smart organizations include formal codes of ethics, which comprise statements of prescribed and proscribed values or behaviors, and thus provide a strategic tool within organizations to inculcate and to demonstrate ethical standards.

Mutual Forbearance between Partners

Forbearance is refraining from cheating. Cheating may take a weak form—failing to perform a beneficial act for the other party, or a strong form—committing a damaging act. The incentives for forbearance arise from the possibility of reciprocity, leading to mutual forbearance. Parties that are observed to forbear may gain a reputation for this behavior, which makes them potentially attractive partners for others. The parties to a successful agreement may develop a commitment to mutual forbearance, which cements the partnership, and, in this way, mutual trust is created, which alters the preferences of the parties toward a mutually cooperative mode. Thus, short-term, self-interested behavior becomes converted to cooperative, trusting behavior.

Competence of Partners

In a smart organization partners will be more willing to share knowledge when they trust in others' competence. It is only natural that they would want to converse with others who have the knowledge and skills regarding the topic at hand, since smart organizations almost always center around a common theme.

Effective Communication and Interaction between Partners

Through communicating with people, we calibrate them, we get a better sense of them and we understand their priorities. Members of a smart organization can therefore increase the trust they are giving and the amount they will trust others, by actively seeking opportunities to communicate with other members.

Trust Maintenance

Besides building trust, it is essential to maintain trust among partners within smart organizations. For instance, through the Smart Bridge project the temporal quality of trust was revealed (Grabowski & Roberts, 1998). Initially, when members in a networked organization were excited about the opportunities joint software development and integration posed, much proprietary product and planning information, as well as member-confidential development and integration strategies were exchanged. Over time, however, as the smart organization matured, and the strength of the linkages between some members faded, the initial trust between some members faded. Some members, although partners on paper, were excluded from planning and integration discussions late in the project, and less proprietary information overall was exchanged as the project matured. Thus, as this networked organization matured, trust among some members waned, suggesting that management of trust in smart organizations requires at least as much effort and interest as management of the smart organization and its linkages. Meyerson, Weick, and Kramer(1996) developed the concept of swift trust for temporary teams whose existence, like partners in a smart organization, is formed around a common task with a finite life span. Such teams consist of members with diverse skills, with a limited history of working together, and with little prospect of working together again in the future. The tight deadlines under which these teams work leave little time for trust building. Because the time pressure hinders the ability of team members to develop expectations of others based on first hand information, members import expectations of trust from other settings with which they are familiar. Analogous to the hyperpersonal model (Walther, 1997), individuals in temporary groups make initial use of category-driven information processing to form stereotypical impressions of others. After the team has begun to interact, trust is maintained by a "highly active, proactive, enthusiastic, generative style of action" (Meyerson et al., 1996).

High levels of action have also been shown to be associated with high performing teams (Iacono & Weisband, 1997). Action strengthens trust in a self-fulfilling fashion: Action will maintain members' confidence that the team is able to manage the uncertainty, risk, and points of vulnerability, yet the conveyance of action has as a requisite the communication of individual activities. In summary, whereas traditional conceptualizations of trust are based strongly on interpersonal relationships, swift trust deemphasizes the interpersonal dimensions and is based initially on broad categorical social structures and later on action. Since members initially import trust rather than develop trust, trust might attain its zenith at the project inception (Meyerson et al., 1996).

Another effective way of holding partners together in a smart organization is through execution of business projects. In a smart organization, projects will need to be renegotiated more quickly and individual partners need the ability to be creative about ways of completing projects based on the principles of trust. The individual partner in a smart organization now has to go beyond basic task cooperation and legal arrangements typical of traditional patterns of hierarchical work organization toward a search for greater trust and autonomy. Trust allows partners to take risks with themselves by being able to make mutual disclosures to others and to develop new task execution styles. Norms of behavior can only work if people know what to expect, but expectations also require some prior commitment, which is based on belief. When rapid change occurs, existing role structures can be destroyed, but without being replaced immediately with legitimate alternative structures. However, people will resist getting rid of norms that have meaning for them, particularly where they think they are already well accepted and approved.

Changing from one situation to another will take time in order for individuals to learn at an emotional level how to come to terms with loss and growth. People who have already invested time and energy in learning how to manage a situation may find change unwelcome when having to learn new roles. The creation of roles can be seen in the life cycle of a networked organization and is believed to go through the stages of forming, storming, norming, performing, and adjourning (Handy, 1995). Trust is vital for commitment during the norming phase. The storming phase is anomic in the sense that people are challenging each other for authority over what is true, correct, and what they want. Order is reached when people accept norms as legitimate according to some criteria and then they can perform. There should be a consensus over norms, but there is a possibility of a minority coercing the others and then legitimacy is weak. Sufficient time must be allowed for people to work through the process using open communications, otherwise they cannot be committed, particularly if there is a requirement for a major change in norms, as one would expect when both planned and unplanned changes occur frequently.

One should also consider how cooperation can be affected by smart organization partnership. Cooperation is a special form of behavior which is concerned with reciprocal actions which are to each other's mutual benefit. Selfish behavior can in a strategic sense lead to cooperation under certain conditions where both parties are mutually dependant on each other for an uncertain period ahead. At an individual level, in one off encounters, people can afford to be selfish (using game-like strategies) because they may not meet each other again. However, where they are likely to encounter the same people again on a repetitive basis the need for cooperation increases because the risk of future retaliation may rise. If trust does exist between people there can still be dissent but this will not involve conflict, which is at each other's expense. In the case of selfish behavior, trust in the other person is not required and behavior may become competitive, as soon as people realize that they are no longer dependent on each other. Alternatively, role relationships may exist which are exploitative because of unequal power and low levels of trust.

Summing up, failure to manage trust properly in a smart organization can be a source of failure in business today. In the following sections the need for trust-based evaluations in smart organizations will be examined.

Security Problems in Ad Hoc Networks

Nowadays, the great flexibility of TCP/IP has led to its worldwide acceptance as the basic Internet and intranet communications protocol. At the same time, the fact that TCP/IP allows information to pass through intermediate computers makes it possible for a third party to interfere with communications in any of the following ways:

- **Eavesdropping:** Information remains intact, but its privacy is compromised. For example, someone could learn your credit card number, record a sensitive conversation, or intercept classified information.

- **Tampering:** Information in transit is changed or replaced and then sent on to the recipient. For example, someone could alter an order for goods or change a person's resume.
- **Impersonation:** Information passes to a person who poses as the intended recipient. Impersonation can take two forms: spoofing or mis-representation.

In the case of smart organizations, many sensitive personal and business communications over the Internet require precautions that address the threats mentioned above. Two most common approaches used to deal with the security problems are encryption and steganography (Mezgar & Kincses, 2003):

- **Encryption:** Transforming the message to a ciphertext such that an enemy who monitors the ciphertext cannot determine the message sent. The legitimate receiver possesses a secret decryption key that allows him to reverse the encryption transformation and retrieve the message. The sender may have used the same key to encrypt the message (with symmetric encryption schemes) or used a different, but related key (with public key schemes).
- **Steganography:** The art of hiding a secret message within a larger one in such a way that the opponent cannot discern the presence or contents of the hidden message. For example, a message might be hidden within a picture by changing the low-order pixel bits to be the message bits.

The increasing use of wireless ad hoc networks such as the Smart Sensor Networks pose challenges to security (Zhou et al., 1999; Zhang, 2000; Deng et al., 2002). First of all, the use of wireless links renders an ad hoc network susceptible to link attacks ranging from passive eavesdropping to active interfering. Unlike fixed hardwired networks with physical defense at firewalls and gateways, attacks on an ad hoc network can come from all directions and target any node. Damage includes leaking secret information, interfering message, and impersonating nodes, thus violating the basic security requirements. All these mean that every node must be prepared for encounter with an adversary directly or indirectly.

Secondly, autonomous nodes in an ad hoc network have inadequate physical protection, and are therefore more easily captured, compromised, and hijacked. Malicious attacks could be launched from both outside and inside the network. Because it is difficult to track down a particular mobile node in a large scale of ad hoc network, attacks from a compromised node are more dangerous and much harder to detect. All these indicate that any node must be prepared to operate in a mode that should not immediately trust any partner.

Thirdly, any security solution with static configuration would not be sufficient because of the dynamic topology of the networks. In order to achieve high availability, distributed architecture without central entities should be applied. This is because introducing any central entity into a security solution may cause a fatal attack on the entire network once the centralized entity is compromised. Generally, decision-making in the ad hoc networks is decentralized and many ad hoc network algorithms rely on the cooperation of all nodes or partial nodes. But new types of attacks can be designed to break the cooperative algorithm. Malicious nodes could simply block or modify the data traffic traversing them by refusing the cooperation or hacking the cooperation. As can be seen from the above, no matter what security measures are deployed, there is always somc vulncrability that can be exploited to break in.

It seems difficult to provide a general security solution for the ad hoc networks. A traditional cryptographic solution is not adapted for the new paradigm of the networks. As can be seen from the above analysis, what is lacked in the ad hoc networks is trust, since each node must not trust any other node immediately. If the trust relationship among the network nodes is available for every node, it will be much easier to select a proper security measure to establish the required protection. It will be wiser to avoid the untrusted nodes as routers. Moreover, it will be more sensible to reject or ignore hostile service requests. Trust-based evaluation thus becomes a before-security issue in the ad hoc networks.

Selection of Partners in Smart Organizations

In the new economy, knowledge is increasingly seen as central to the success of organizations and an asset that needs to be managed. The ability of an enterprise to manage knowledge as an asset (and provide a good return on investment) is seen as the key to survival in a global business environment in which the efficiencies of mass production of commodity goods have been successfully exported to emerging economies. The major issue of knowledge management is to place knowledge under management remit to get value from it—to realize intellectual capital. That intellectual capital can be regarded as a major determinant of the difference between a company's book price and the total value of its physical assets. For a successful company, this difference can be considerable, representing the difference between the way the company is seen by financial experts and by the market. For example, there is a great difference between the book price and share value of recently-launched biotechnology companies, whose market value is clearly based on their knowledge assets, rather than traditional capital. However, while the world of business is experienced in managing physical and financial capital, smart organizations have difficulty finding solutions to practical questions concerning partnership management, such as:

We are involved in an exciting project with four other companies. How can we tell whether all these partners would collaborate?

Market needs change often these days and we are always bringing new partners into projects. How can we select the right partners?

The main issue is that partners of the smart organizations are quite intractable from a knowledge management point of view. By their very nature such partners create a great deal of new knowledge, which as such is of high value to the smart organization. However, the knowledge of how and why they created, and what they had created is not clear since it involves the interactions among a group of different people. Since smart organization partners do not have frequent face-to-face interaction, the core company has to have total faith that the partners will do the job they are assigned, and do it right. This leaves core companies with the daunting task of selecting partners who are not only able to work on their own, but can also function within a team structure—self-motivated partners. When it works, the organization processes flow nicely. However, when one partner starts slacking, the ramifications are dire in consequence. A frequent occurrence is when a partner joins the smart organization with certain expectations, but those expectations are modified by an unexpected technology breakthrough. Suppose a regional airline was brought into a certain route-alliance smart organization to harmonized express, general air cargo and heavy express air cargo products in the Asia Pacific Region, and it is just found that there is a growing need for the smart organization to focus on harmonizing dangerous good products. Assuming that due to national aviation legislations the partner airline cannot handle this kind of product, it makes sense for the core air cargo organization to switch partners. And suppose that the original airline has invested a considerable amount of capital in developing general and heavy cargo products handling equipment for the core organization, in addition to carrying out research on European air cargo markets. In the traditional business system, each partner suboptimizes for its own goals, which in this case would provide a less than optimal service, express general and heavy cargo handling only, leaving the dangerous goods handling to the competitors. A better business system will agilely adjust to the customers' need and at the same time reward the general cargo-handling partner for essentially putting itself out of the smart organization. In such a case, that partner has to have trust that the smart organization will deal with it fairly whether it is fully in the smart organization or out. And the smart organization has to have trust in each partner that they will strive to optimize the enterprise even when it reduces or eliminates its own role.

While trust has long been a major issue in the organizational literature, there is little agreement on a single model of trust that applies to all partner evaluation contexts. One can only see from the literature that although trust is pivotal in reducing the high levels of uncertainty endemic to the global and technologically-based environment, interfirm trust was rarely considered in the evaluation of partners in a smart organization. For effective smart organization management, it is clear that electronic handshake would need to be based on trust between partner companies as well as the correct protocol. To facilitate the evaluation of interfirm trust a partner evaluation decision support system is suggested below (Lau & Wong, 2001; Wong & Lau, 2003).

Neural Data Mining System (NDMS)

In larger organizations, many different types of users with varied needs utilize the same massive data warehouse to retrieve the right information for the right purpose. Whilst data warehouse is referred to as a very large repository of historical data pertaining to an organization, data mining is more concerned with the collection, management, and distribution of organized data in an effective way. The nature of a data warehouse includes integrated data, detailed and summarized data, historical data, and metadata. Integrated data enable the data miner to easily and quickly look across vistas of data. Detailed data is important when the data miner wishes to examine data in its most detailed form, while historical data is essential because important information nuggets are hidden in this type of data. Online analytical processing (OLAP), an example of architectural extension of the data warehouse, has become popular in the last couple of years. OLAP refers to the technique of performing complex analysis over the information stored in a data warehouse. For instance, by using the OLAP Hub (SPSS, 2004), a commercial Web-based online analytical processing software, the managers within a smart organization can access and analyze information about organizational performance, share their findings with others immediately via the Web, and take action as opportunities arise. Consequently, more informed decisions will result and this in turn will lead to enhanced productivity and increased profitability.

Once a data warehouse is set up, the attention is usually switched to the area of data mining, which aims to extract new and meaningful information. In other words, a pool of "useful information" that has been stored in a company data warehouse becomes "intelligent information," thereby allowing decision-makers to learn as much as they can from their valuable data assets. In this respect, neural networks can be deployed to enhance the intelligence level of the OLAP application. Neural networks search for hidden relationships, patterns, correlations, and interdependencies in large databases that traditional information gathering methods (such as report creation and user querying) may have overlooked. The responsibility of the neural network is to provide the desired change of parameters based on what the network has been trained on. Intrinsically, a sufficient amount of data sample is a key factor in order to obtain accurate feedback from the trained network. As neural network is capable of learning relationships between data sets by simply having sample data represented to their input and output layers (Herrmann, 1995), the training of the network with input and output layers mapped to relevant realistic values with the purpose to develop the correlation between these two groups of data will not, in principle, contradict the basic principle of neural network. With a trained network available, it is possible that recommended action can be obtained with the purpose to rectify some hidden problems, should that occur at a later stage. Therefore, in the training process of the neural network, the nodes of the input layer of the neural network represent the data from the OLAP and those of the output layer represent the predictions and extrapolations. It should be noted that the output information from the OLAP could be used to refine the OLAP data cube so as to continually update the database over time.

Characteristics of NDMS

The data interchange within the NDMS encompasses three modules, namely OLAP module, data conversion (DC) module, and neural network (NN) module (Figure 1). The data repository, which aims to support efficient data interchange among the three modules, is essential for the coordination and updating of information from various sources. As for the OLAP module, it consists of descriptive data (dimensions) and quantitative value (measures), both of which generate the OLAP data cube by building up two elements, namely, fact table and dimension (Erik, George, & Dick, 1999). In the fact table, the required data and user-defined methods for analysis are specified clearly. In the descriptive data of OLAP, the different dimension levels are defined for further computational use on different views of OLAP data cube. Typical dimension includes location, company, and time; whereas typical measure includes price, sales, and profit. With a multidimensional view of data, the OLAP module provides the foundation for analytical processing through flexible access to information. In particular, this distinct feature can be used to compute a complex query and analyze data on reports, thereby achieving the viewing of data in different dimensions in an easier and more efficient way. To illustrate the benefits of OLAP as compared to the traditional approach of data management using structured query language (SQL), an example is shown here to benchmark their underlying methodologies of associated operations.

In the SQL approach, when a user needs to retrieve information across multitables, users must clearly define the necessary tables for finding the specific information. For instance, when a user needs to know how much sales is taken for a certain year and city and uses the Internet as promotion media, the tables and their relationships must be clearly defined. In normal practice, a query command line written in SQL as shown below can be used to retrieve information from the tables:

"Select sum(a.store_sales) from sales_fact_2004 a, promotion b, region c, store d,

where a.store_id = d.store_id and c.region_id = d.region_id and

b.promotion_id = a.promotion_id and c.sales_country= China and b.media_type = Internet"

Figure 1. Characteristics of NDMS

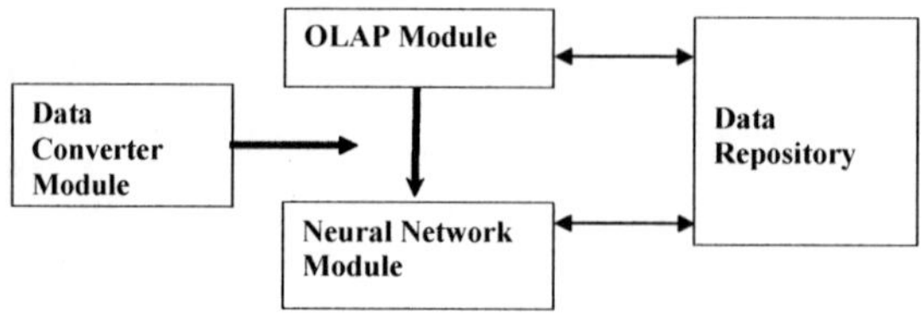

When using the OLAP module, the table used for the query and the data used to perform the calculation are defined separately. Then, the user builds up a complex calculation on individual members to meet the specific requirements. Because the calculation and analysis have been precomputed in the OLAP server previously, only a simple Multi-Dimension eXpression (MDX) is necessary to construct for retrieving identical result as shown in the following:

"Select [Measures].[Store Sales] on columns, [Store].[Sales_country] on rows from sales where ([Promotion].[Media_Type].[Internet], [Region].[Sales_country].[China])"

In the above expression, it can be shown that MDX is simpler and clearer than SQL statement. When the user requirement is changed, there is only a small part of the OLAP data cube needed for alteration in order to fulfill the user requirement. In this respect, some minimal change of the MDX is necessary. Comparing with the traditional approach, the SQL statement needs to be rewritten to meet the new requirement.

In general, MDX is suitable for creating decision support functionality, and a typical example has been provided as follows:

*"IF ([Measures].[Unit_Sales] > 1000, [Measures].[Store_Sales] * 0.8, [Measures].[Store_Sales])"*

Before the implementation of OLAP, the calculated member is constructed by the measures. In this case, in an example that aims to find a suitable business partner for a particular task, the method has been depicted in Figure 2.

Figure 2. Partner score method

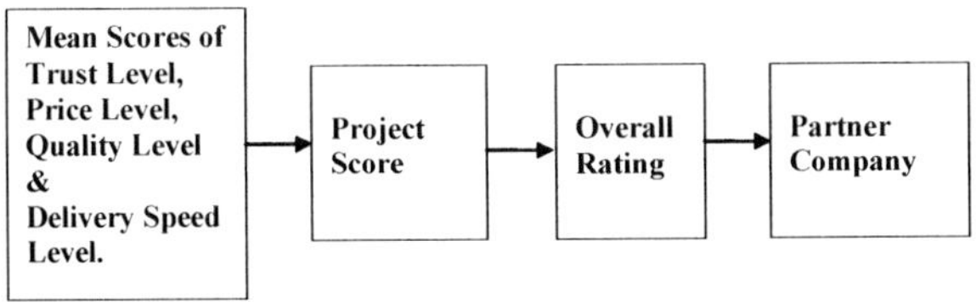

Evaluation of SO Partners

Industries in the 21st century find themselves in a world increasingly globalized and concerned with speed, quality, and cost. As a result, many enterprises now manufacture and distribute globally. The evaluation of SO partners generally involves distinct types of decisions. For instance, in the evaluation of product distributors, one focuses on a partner to handle a company's product in the foreign or domestic market. Hence, access to markets and market intelligence are likely to be of key importance, together with reliability and control over conditions of distribution and sales. On the other hand, evaluation of supply partners involves the identification of a partner for meeting the company production requirements. In this case, supply chain variables such as material quality, price, supply lead time, and inventory control are generally considered to be of key importance. In both cases, trust between top management teams of the core and partner companies is of fundamental importance.

The selection of attributes or variables for consideration in the comparison of smart organization partners is quite important to the eventual outcome. For instance, considerations relating to financial objectives, expected quality level, safety standards, problem resolution mechanisms, and so forth. In addition, intangible benefits and risks are often present. The set of attributes should include all the criteria that are needed in order to describe the differences between partner competencies. It is desirable that the list is complete, but if there is a criterion on which all the partner competencies are equal, then that criterion can be discarded, since it will not have any differential impact on the final partner rankings.

In Figure 2, the scores of trust level, price level, quality level, and delivery speed level for each partner are used. The mean score of a project is calculated by taking the mean scores of the trust level, price level, quality level, and delivery speed level of the previous projects that have been carried out by the partner company concerned. Finally, the overall average of the partner company is determined by accumulating the prespecified weighting of the latest projects. The overall average can then be assessed by the top management of the core company. Since the OLAP technology is a user-friendly and software-independent tool, it can be embedded in most client server development tools and Web development tools. In the following case example, after the customer submits the requirement to the NDMS server, an MDX query can then be executed in order to retrieve available service providers based on the core competence specified.

With the OLAP module as a front-end component, the neural network (NN) module is employed as a back-end part of the NDMS, which concentrates on providing essential information such as alertness of abnormal scenarios based on the pattern of historical data. However, since the output data from the OLAP data cube may not be able to be used directly by the NN module due to possible data incompatibility, it is essential that a data translation mechanism (DTM) is to be incorporated to act as a "bridge" to link the two modules together to form an integrated unit.

The DTM module is meant to play this important role. In brief, the DTM module concentrates on achieving efficient data transfer between the OLAP module and the NN module, which requires specific data format for mapping the input nodes to guarantee proper operation. With the formatted data available via the DC module, the focus is now turned to the NN module, which aims to set up a suitable network topology in order to identify any correlation within the data pool. The NN module is meant to project possible outcomes—good or bad—based on the available pattern of data, thereby alerting users of detected abnormal behavior in terms of company performance and other hidden business issues. This provides important advice to support critical decision-making in the SO. In the following case example, parameters such as trust between management teams, product quality, and product cost are abstracted from recent company performance records. With the assistance of an expert team, many companies' past behavior based on the selected parameters can be classified and ranked. In general, for the setup of a neural network, a number of inputs are required to enable such network to take into account the various multiple factors that may influence the performance assessment of a particular company. In this research, the neural network consists of four input nodes (five sets of the last five records including trust, quality, cost and speed) and five output nodes (various suggested actions to be taken), as shown in Table 1. To achieve the objective of producing a reliable "trained" neural network, statistical data have to be mapped to the input and output nodes of the neural network. In this respect, it is recommended that at least 100 sets of data are required to train the network in order to develop a reliable module for the NDMS. With the availability of a fair amount of data sets, the next step is to train a neural network.

This means the parameters include network construction, and training data files are needed for definition. Then, users can apply analysis tools to provide insight into how the network is to be trained, as well as the appropriate topology of the network for the specific purpose. It is also likely that users will need to fine-tune the training parameters such as iteration number, learn rate, momentum, and so forth (Qnet, 2000), so that the specified values match the training characteristics for the model. After the completion of the training process, such neural networks can then be used as a knowledge repository to monitor companies' performances, and provide decision support to users who then consider necessary actions to be taken.

After the training process, the trained neural network can then be recalled in order to process the new inputs through the network. In order to describe in more detail about the recall operation under the NN module, five latest track records of a company are required to be mapped to the input nodes for analysis. Output data from the NN module will predict the company's performance based on the configuration of the trained network. A case example in the next section will elaborate how this works.

Case Illustration

To verify the applicability of the NDMS, a prototype system has been developed, based on the framework of the NDMS as proposed in this paper. Pursuing the NDMS infrastructure that has been defined in the previous section, the OLAP module has generated a pool of useful data and accordingly, the NN module has created a reliably trained neural network. Next, five latest track records of a company have been gathered and listed as follows. In this case four factors—trust between top management, quality, cost and delivery speed—are being considered, and performance scores (PS) ranging from 1 (least point) to 7 (highest point) are used to assess the partner company, as shown in Table 1.

Table 1. Five latest records of Company A

Company A	Performance Score(PS)			
	Trust between Top Management	Product Quality	Product Cost	Delivery Speed
Latest record	6	5.6	6.5	6.2
2nd latest record	6.4	4.8	5.9	6.4
3rd latest record	5.9	4.1	6.0	5.7
4th latest record	6.6	3.8	5.6	5.8
5th latest record	6.3	3.5	5.5	6.0

Table 2. Output from the NN module

Company A	Output from NN module
Price quoted reflects current market situation	1
Product quality has been compromised to meet the quoted price	1
Dependability of company	1
Potentially competent	0.5
Additional assessment of company performance is required	1

After such information has been input, the NN module gives an assessment report back to the user, thus supporting the user to take action if deemed necessary. In the following table, "0" output from the NN node indicates a negative suggestion to the associated statement, and "1" is the positive suggestion, whereas "0.5" indicates that there is not enough data to justify a firm suggestion.

Based on the NN output results as shown in the table, it can be seen that although Company A is trustworthy, it seems to have problem in meeting the agreed quality level, and it is suggested that additional assessment regarding the company's performance is needed. Based on the suggestion of this assessment report, Company A was approached in order to find out the reasons behind the abrupt change in product quality. After an organized investigation of the issue, it was found that two of the key staff responsible for the quality assurance department left the company several months ago to join a newly established multinational corporation. Because of this unexpected change, the company suffered an unprecedented "brain-drain," resulting in the sudden decline of quality level of certain mainstream products.

Because of the situation, Company A has been advised to implement some best practices related to quality assurance. In this case, the concept of total quality management (TQM) has been adopted and a TQM consultant team has also been requested to facilitate the implementation of such practice in the company. At this stage, it is still difficult to tell whether the company could significantly reverse the downturn performance in terms of product quality. However, because of the signal generated from the NDMS, the problem of a business partner has been revealed, and a prompt business decision could be made with supporting assessment report, thus avoiding the loss of a trusted business partner, which can in turn weaken the overall performance of the smart organization. This case example indicates that the introduction of the NN module to the OLAP module is able to significantly upgrade the decision support functionality of the smart organization. However, the results obtained, so far, are by no means perfect although they demonstrate that the suggested NDMS is viable.

Limitations and Recommendations for Further Research

This study mainly focused on the practical need to consider trust between partner companies in a smart organization. Owing to the circumstances pertaining to the schedule of the study, the perception of a limited number of managers responsible for partner selection was used to verify the importance of interfirm trust. It would be desirable to extend the analysis to all the dimensions pertaining to trust, and explore the effect of trust on the performance of partner companies. Additionally, qualitative studies on mapping the cognitive processes of senior managers responsible for evaluation of partner companies in the smart organization would also be helpful. Here conjoint analysis might be applied. In the present context, conjoint analysis is concerned with understanding how senior managers make choices among partner companies, so that a smart organization can choose new partners that better serve its needs. Although it has only been a mainstream research technique for the last decade or so, conjoint analysis has been found to be an extremely powerful of way of capturing what really drives top managers to select one partner over another, and what the top managers of a particular smart organization really value. A key benefit of conjoint analysis is the ability to produce dynamic decision models that enable companies to test out what variables they would need to consider to improve the decision-making process. Coupled with the decision support tool proposed, such a methodology would help to understand how managers actually make such decisions and use experience in partner evaluation.

Conclusion

In order to achieve the target mission of a smart organization, the core company must be clear about its business aims and objectives. It must assemble a set of partner companies that can deliver to those objectives. It must support them in doing so and trust them to do so. In providing security solution to ad hoc networks or in the selection of partners, a company has to analyze the trustworthiness of the potential partners, their capabilities, track record, and future potential. However, there is no single model of trust that could be applied in practical situations. Hence, there is great pressure placed on the CEOs as the decision makers for selection of satisfactory partner companies. In this chapter, an intelligent decision support system for smart organization partner evaluation is introduced. It demonstrates the benefits of using a combination of technologies to form an integrated system which capitalizes on the merits and at the same time offsets the pitfalls of the involved technologies. A special feature is that the trust between management teams of companies could be incorporated as one of the evaluation criteria. The NDMS has been found to be feasible in predicting the problems of companies as shown in a case example described in the paper. As suggested, further investigations on (i) key dimensions relating to trust between companies within a smart organization and (ii) the impact of trust on overall performance of partner companies would be needed.

Acknowledgment

The author wishes to thank the Department of Mechanical Engineering, The Hong Kong Polytechnic University for supporting this study.

References

Augustine, N. R. (1997). Reshaping an industry: Lockheed Martin's survival story. *Harvard Business Review, 75,* 83-94.

Axelrod, R. (1990). *The evolution of co-operation.* Penguin Books.

Bhimani, A. (1996) Securing the commercial Internet, *Communications of the ACM, 39*(6), 29-35

Business Week (1993). The virtual corporation. February 8, 98-103.

Coyle, J., & Schnarr, N. (1995). The soft-side challenge of the "virtual corporation." *Human Resource Planning, 18,* 41-42.

Davidow, W. H., & Malone, W. S. (1992). *The virtual corporation.* New York: Harper Collins Publishers.

DeSanctis, G., & Poole, M. S. (1997). Transitions in teamwork in new organizational forms. *Advances in Group Processes, 14,* 157-176.

Erik, T., George, S., & Dick, C. (1999). *Microsoft OLAP solutions.* New York: John Wiley & Sons.

Filos, E., & Banahan, E. (2000) Will the organization disappear? The challenges of the new economy and future perspectives. In L. M. Camarinha-Matos, H. Afsarmanesh, & R. J. Rabelo (Eds.), *E-business & virtual enterprises* (pp. 3-20). Dordrecht: Kluwer.

Gamble, P. R. (1992). The virtual corporation: An IT challenge. *Logistics Information Management, 5*(4), 34-37.

Giddens, A. (1993). The nature of modernity. In P. Cassell (Ed.), *The Giddens reader* (pp. 284-316). Stanford University Press.

Handy, C. (1995). Trust and the virtual organization. *Harvard Business Review, 73*(3), 40-50.

Herrmann, C. S. (1995). A hybrid fuzzy-neural expert system for diagnosis. *Proceedings of International Joint Conference on Artificial Intelligence* (pp. 494-500).

Humphrey, J., & Schmitz, H. (1998). Trust and inter-firm relations in developing and transition economies. *The Journal of Development Studies, 34*(4), 32-61.

Iacono, C. S., & Weisband, S. (1997). Developing trust in virtual teams. *Proceedings of the Hawaii International Conference on Systems Sciences*, Hawaii. (CD-ROM).

Jarvenpaa, S. L., & Leidner, D. E. (1998) Communication and trust in global virtual teams. *Journal of Computer-Mediated Communication, 3*(4).

KPMG. (1999). *Electronic commerce—The future is here.* Norton and Norton Institute.

Kramer, R. M., & Tyler, T. R. (Eds.). *Trust in organizations: Frontiers of theory and research* (pp. 166-195). Thousand Oaks, CA: Sage.

Lau, H., Chin, K. S., Pun, K. F., & Ning, A. (2000). Decision supporting functionality in a virtual enterprise network. *Expert Systems with Applications, 19*(4), 261-270.

Lau, H., & Wong, T. T. (2001). Partner selection and information infrastructure of a virtual enterprise network. *Computer Integrated Manufacturing, 14*(2), 186-195.

Lea, M., & Spears, R. (1992). Paralanguage and social perception in computer-mediated communication. *Journal of Organizational Computing, 2*(3/4), 321-341.

Lipnack, J., & Stamps, J. (1997). *Virtual teams: Reaching across space, time, and organizations with technology.* New York: John Wiley & Sons.

Lippitt, G., & Lippitt, R. (1978). *The consulting process in action.* University Associates Inc.

Lorenz, E. (1999). Trust, contract and economic cooperation. *Cambridge Journal of Economics, 23,* 301-315.

Marris, P. (1974). *Loss and change, institute for community studies.* Routledge and Kegan Paul.

Maskell, P. (2000). Social capital, innovation, and competitiveness. In S. Baron, J. Field, & T. Schuller, (Eds.), *Social capital: Critical perspectives.* New York: Oxford University Press.

Mayer, R. C., Davis, J. H., & Schoorman, F. D. (1995). An integrative model of organizational trust. *Academy of Management Review, 20,* 709-734.

Mezgar, I., & Kincses, Z. (2003). The role of trust in information technology management. In A. Gunasekaran, O. Khalil, & S. M. Rahman (Eds.), *Knowledge and information technology management—Human and social perspectives* (pp. 283-304). Hershey, PA: Idea Group Publishing.

Meyerson, D., Weick, K. E., & Kramer, R. M. (1996). Swift trust and temporary systems. In R. M. Kramer & T. R. Tyler (Eds.), *Trust in organizations* (pp. 166-195). Thousand Oaks, CA: Sage.

Moore, M. (1999). Truth, trust and market transactions: What do we know? *The Journal of Development Studies, 36*(1), 74-88.

Musthaler, L. (1995). Effective teamwork virtually guaranteed. *Network World*, October 16, SS10-SS11.

Parkhe, A. (1998). Understanding trust in international alliances. *Journal of World Business, 33*(3), 219-240.

Qnet v2000 (2000). Retrieved March 23, 2005, from http://qnetv2k.com/.

Raman, D. (1996). Cyber assisted business - EDI as the backbone of electronic commerce, EDI-TIE.

SmartBridge. (1997). Retrieved on March 23, 2005, from http://www.hokie.bs1.prc.com/maritech/dsc94-44.htm.

Walther, J. B. (1997). Group and interpersonal effects in international computermediated collaboration. *Human Communication Research, 23*(3), 342-369.

Wong, T. T., & Henry, C.W. Lau (2003), The impact of trust in virtual enterprises. In A. Gunasekaran, O. Khalil, & S. M. Rahman (Eds.), *Knowledge and information technology management—Human and social perspectives* (pp. 153-168). Hershey, PA: Idea Group Publishing.

Section IV

Communication and Security Technologies for Smart Organizations

Chapter VII

New Challenges for Smart Organizations: Demands for Mobility – Wireless Communication Technologies

István Mezgár, Hungarian Academy of Sciences and
Budapest University of Technology, Hungary

Abstract

The chapter introduces the different types of wireless technologies that can be applied in smart organizations. Smart organization (SO) is an outstanding representative of networked organizations, as its organization structure, communication, and knowledge-based applications are coordinated and all networked. The chapter describes the communication demands of SO, taking care on wired and especially wireless networks that offer mobility for users. Access at anytime from anywhere to enterprise information for registered users guarantees mobility that is a basic demand for a dynamic organization today. Security, trust, and interoperability aspects are also discussed as important characteristics of the up-to-date infocom systems. Finally, the main impacts of wireless technologies on smart organizations are summarized. Through the survey of the structure and operation of wireless technologies and their impacts, it is easy to understand that wireless communication technology has a strategic role in the effective, competitive operation of networked organizations.

Introduction

Today, the developments of information technology, telecommunication, and consumer electronics are extremely fast. The ability of different network platforms to carry similar kinds of services and the coming together of consumer devices such as the telephone and personal computer is called "technology convergence." The ICT (information and communication technology), the "infocom" technology, covers the fields of telecommunication, informatics, broadcasting, and e-media. A very fast-developing field of telecommunication, wireless communication gets a growing role in many fields as well. The connection of mobile devices to the Internet established basically new possibilities and services for the users.

Based on the results of the information and communications technologies, a new digital economy is arising, as the result of this technological convergence is not just about technology. It is also about services and about new ways of doing business and of interacting within the society. The impact of the new services resulting from convergence can be felt in the economy and in the society as a whole, as well as in the relevant sectors themselves. Because of this great impact of information, technologies, and the level of knowledge content in products and services, the society of the 21st century is called the Information and Knowledge Society (Ungson, 1999).

This new economy needs a new set of rules and values, which determine the behavior of its actors. Participants in the digital market realize that traditional attitudes and perspectives in doing business need to be redefined. In this dynamic and turbulent environment that requires flexible and fast responses to changing business needs, organizations have to respond by adopting decentralized, team-based, and distributed structures variously described in the literature as virtual-, networked-, cluster-, and *smart* organizations. One main aspect of these approaches is that organizations in this environment are networked; that is, interlinked on various levels through the use of different networking technologies.

Since the base of networked organizations are the interdependent, separate production and service teams and units, cooperation and collaboration among them is of vital importance. The structure, the communication systems and the collaborating people, teams, and organizations that define today's organizations characteristics must be harmonized to accomplish complex, demanding tasks. The collaboration means contacts among users, so human beings have outstanding importance. A very important element of this human contact is trust, so in a networked organization, trust is the atmosphere, the medium in which actors are moving.

The collaboration is done through different media according to the actual demands of the tasks. A new way of connection is the application of different wireless technologies for communication in networked organizations where interoperability is an additional important factor. Wireless technology means mobility, namely individuals are available independently from location and time. This mobility is an important attribute of today's organizations.

The chapter introduces the basic wired and wireless communication possibilities for smart organizations and some connected techniques, and technologies as well. As trust is the base for all cooperation and collaboration, its role in smart organization is also presented. Security is an important component in trust building, so security services and their application in the different networks are introduced as well.

Smart Organizations and ICT

The Smart Organization

The digital economy offers a market environment for participants with the following characteristics and demands:

- Networking and horizontal communication, including the smart product,
- Networked environment,
- Knowledge-based technologies,

- Simplification and coordination of structure,
- Customer focus and real-time, ubiquitous responsiveness to technical and market trends (what customers want, anytime, anywhere),
- Flexibility, adaptability, agility, mobility,
- Organizational extendibility, virtuality,
- Shared values, trust, confidence, transparency, and integrity,
- Ability to operate globally, cooperating with local cultures.

In this turbulent environment only those organizations can survive which effectively apply the results of the different disciplines. Smart organization (SO) belongs to this kind of category. A definition of SO is as follows: "The term "smart organization" is used for organizations that are knowledge-driven, internetworked, dynamically adaptive to new organizational forms and practices, learning as well as agile in their ability to create and exploit the opportunities offered by the new economy"(Filos & Banahan, 2000).

There are three characteristics of smart organizations that make them really special:

- They are motivated to build collaborative partnerships, which encourage and promote the discussion of ideas. Customer focus and meeting customer expectations is recognized as a key success factor.
- Smart organizations can respond positively and adequately to change and uncertainty, so they survive and prosper in the new economy.
- Smart organizations can identify and exploit new opportunities through applying the strength of "smart" resources; that is, information, knowledge, relationships, and innovative and collaborative intelligence.

Cooperation and communication are the basic technologies for SO operation. In the following sections the important technologies, services and methods connected to communication technologies will be discussed trust, security and their applications.

Besides the Internet, new (or pilot phase) solutions are offered: powerline communication (using the electric power grid), and as an efficient extension of the Internet the Grid technology and different types of wireless networks.

Demand of Mobility

Mobility means availability everywhere, the possibility to be connected in the office, in the meeting room, in the car, in a train, in the airport, and even at home. Today, most companies are aware of the importance of employees continuously being connected. Being connected and being able to contact anybody associated with the organization (e.g., customers, colleagues) is very important both for the individual and for the overall efficiency of the organization. The "classic" networked companies aspire to mobilize their workforces through application of wireless communication technologies as mobility brings obvious advantages to the enterprises, including increased productivity, improved communications, cost reductions, and revenue creation.

This mobile connection can be either a voice or a data connection to an enterprise network or to the Internet. Depending on the employee's position inside the organization, the requirements for this connectivity in terms of bandwidth, end user devices, services, and so forth are very different.

Mobility can be addressed with different solutions and technologies; each solution or technology has its own advantages and inconveniences. Mobility can be achieved by using different types of wireless networks such as satellite communication, Wireless Wide Area Networks (WWAN—different types of mobile phone systems—GSM, GPRS, UMTS, iMode), Wireless Metropolitan Area Network (WMAN), Wireless Local Area Networks (WLAN, such as Wi-Fi—also called mobile Internet [IEEE standard 802.11a/b/g]) and Wireless Personal Area (or Pico) Network (WPAN—e.g., Bluetooth, RFID, IrDA2).

The main structural factor for a mobility solution is the radio and access technology deployed. In the last ten years, some technologies have been developed from a technical as well as a business point of view. Two very strong drivers for mobility are now in place at the infrastructure level, one for indoor (WLAN), and another one for outdoor (cellular solutions such as FLASH-OFDM).

For voice transmission, most of these users will want to use the same phone with a single directory, a single voice mailbox, and a common set of services. A single phone using, for example, Voice over IP (VoIP) through a broadband connection when at home or in the office, but switching automatically to a mobile network when on the move.

Beyond the enterprise, consumers are also discovering the advantage of mobility. Phones became mobiles some years ago, and the same trend is currently happening to the computer world, where consumers are shifting from fixed desktops to more convenient laptops. Most users want to be able to communicate on any device, anytime, anywhere, without any configuration limitation. These users want session continuity between various access networks, giving them continuous and seamless connection to the best available network. (e.g., existing Satellite-GSM phones). There are various solutions (some under research still) that enable seamless mobility between different access networks for voice and data connections. These solutions will make the end-user's experience more enjoyable and will open new opportunities for operators and enterprises. With Mobile IP, the enterprise and public operator can provide seamless "always-on" network access to mobile professionals.

Mobility between the different networks (WLAN, WiMAX, UMTS, etc.) can be coordinated at the Internet Protocol (IP) level, where

- no user intervention is required when the access network changes,
- connectivity is not interrupted during the move,
- any application that tolerates packet loss will remain available, allowing ongoing session recovery.

So, main advantages of wireless and mobile communication are that anybody, from anywhere at anytime, can make contacts, which means increased flexibility and time reduction in business processes. Today, some type of mobility is a must for enterprises that want to hold on in the global market competition. According to market researcher Gartner (2003), 45% of the American workforce is using mobile technology of some kind, including laptops, PDAs, and new sensor networks. By 2007 more than 50% of enterprises with more than 1,000 employees will make use of at least five different wireless networking technologies.

Demands of Communication in SO

Communication in SO is very intensive and a great amount of confidential data and information is changed among the partners. The list of general requirements can contain, in some cases, contradictory items as well. The main requirements in case of novel networks are as follows:

- great communication speed,
- shared access to files, data/knowledge bases,
- exchange of picture, voice—multimedia applications,
- online/real-time access,
- access for anybody, from anywhere, at any time—mobility,
- reliable, secure exchange of information,
- intelligent user interfaces,
- easy and cheap installation.

The additional demands for present wireless communication technologies from the side of production systems are:

- security level should be equivalent with wired systems,
- increase data transfer speed,
- decrease operation costs,
- develop wireless communication chain (satellite -> WPAN)

The general communication requirements for an SO can be summarized in the following:

- **Integration of different communication forms and resources:** Communication through connected telephone-, computer-, and cable networks, and application possibilities of different protocols, connecting wired and wireless equipment.
- **Reliable and high quality communication services:** Reliability covers the high on-service time (technical reliability), the high availability (well designed/balanced network—resource reliability), the HW and SW security, both for equipment and communication lines (access reliability), well controlled/organized networks (organization reliability), all with reasonable cost.

- **Global time coordination:** It is essential the exact coordination of the different actions in time, so a "general time" has to be declared for communication.
- **Traceable communication:** Traceability means to document and audit the communication in a way that fulfills the requirements of bookkeeping (e.g., delivery report and receipt notification) and legal aspects (e.g., digital signature).

The security requirements for an SO can be listed as follows:

1. Protection of all types of enterprise data (for all company forming the SO). Privacy and integrity of all types of documents during all phases of storage and communication (Data and communication security—Certification, Encryption),
2. To enable companies confidential access control,
3. Authorization and authentication of services (digital signature).

These services need to be flexible and customized to meet a wide array of security needs, including specific high-level requirements. In order to fulfill the communication and security demands, some basic aspects have to be taken into consideration while selecting security and communication technologies:

1. Platform independent SW tools have to be applied,
2. Standards have to be applied (accepted and "de facto" standards as well),
3. Appropriate architectures with ability to integrate different resources.

Fulfilling all types of the introduced requirements for individual enterprises would be very hard if not impossible, so different general network and organizational structures have been developed that have been carefully designed and tested. These structures can be defined as reference architectures, and they are available both for the organization and for the information infrastructure of networked organizations.

Selection of Communication Network Technologies for SO

In the case of networked organizations, the most important market demands are the time-to-market (throughput time) and flexibility, so decreasing time and increasing flexibility are the main goals for enterprises. In order to shorten the throughput and operation time there are numerous opportunities, in the following some of them are listed:

- speed up communication among departments/individuals/machines,
- optimize (business) processes -> optimize information/material flow-routes,
- make more effective the working process—workflow modifications,
- increase availability—extend effective working time (24/7/365 method),
- increase efficiency of information/knowledge exchange (team-work, co-operation, coordination).

Information and communication technologies play a key role in most of the points of the above items. In the cooperation and coordination technologies communication and its speed, and the availability rate of partners, have basic importance.

In communication the conventional tools are the telephone, fax, and writing letters. On the next level are the computer network-based solutions, such as e-mail, ftp, and telnet. A higher quality of communication media is the WEB-based communication solutions. Through WEB pages a secure, easy and fast communication can be realized.

In case of WWAN, SMS, MMS (multimedia messages), and Mobile Internet (WAP) are the forms of communication. Mobile network solutions are spreading quickly also in enterprise communication. Developers and service providers offer compact mobile phone-based packages for enterprises that makes possible receiving e-mails and browsing company databases from anywhere.

Today, there is no communication technology (data or voice) that would entirely fulfill alone the communication demands of smart organization. The main characteristics listed in the previous subchapter cannot be represented by one network type yet, so there are different approaches that cover partially the demands of the market; networked organizations are using both wired and wireless technologies in an integrated way.

Wired communication remains the base for communication, but in many fields mobile and wireless tools take over the first place. Internet technology (IP) is the base of broadband wireless communication. A smooth integration of wired and wireless, mobile communication technologies can be seen on the market as well. There are some novel wired technologies as Powerline communication (PLC—ease of installation, operation) and the GRID technology (easy information access and handling) that have special characteristics and they are applied parallel with wireless technologies. In the field of voice transfer, VoIP seems to be a ruling technology.

The different wireless networks—Satellite, WWAN, WMAN, WLAN and WPAN—(today) can be partially connected and integrated according to the actual needs, developing very complex and powerful networks for the networked organizations. The new networking technologies extend company data, back-end information systems, and e-mail to mobile employees, broadens the accessibility of mission critical data. Mobile access modifies the way workers interact with colleagues, customers, and suppliers.

Based on the main characteristics of WN the main advantages that wireless communication can offer for DPS are the following:

- mobility—from anywhere,
- availability—at any time,
- flexibility—for anybody,
- easy and fast installation,
- low installation/operation costs,
- competitive (with the wired systems) data exchange speed.
- specialized protocols (e.g., in the case of WPAN)—for sensors and communication.

Comparing the demands of enterprises, firms, and novel production systems with the characteristics of wireless communication systems (that can be/has been already integrated with computer and information systems) it can be stated that wireless networks have an outstanding strategic role in life cycle of enterprises.

Selection Aspects of Wireless Networks

There are different orientations of networked systems approaching production in general: service industry, manufacturing industry, energy industry, and so forth. The applications of the various types of wireless technologies are in these cases extremely diverse. The latest generation of communication technology, the wireless and mobile networking technologies, have an especially great influence on time factor and flexibility that is very critical in today's production, so wireless technology has a strategic role. The aspects that are influencing the application of wireless technologies can be grouped as follows (Lucent, 2003):

- **Need of Users:** The primary driving need for enterprise users to adopt a mobile data service is access. Key horizontal applications that apply to virtually all users are e-mail (with attachments), corporate intranet access, and general Internet access. The always-on, the real-time access is important for field sales and service workers. To be most productive, they require an always-on, high-performance mobile data solution that can be relied upon wherever they need it. Coverage is critical, and adequate speed for their key applications is very important. In order to define a solution that can provide coverage to users wherever they go, it is necessary to understand where mobile workers travel throughout the course of their work day, and what they need to do in different locations.
- **High-Level Security:** applications, speed, and coverage are not enough. Solutions must address corporate security standards for the enterprise. The corporate IT department must be assured that the user's information will be secure and that company PCs, servers, and data networks cannot be compromised through the mobile data network by unauthorized intruders. Ease of use and cost control are also corporate concerns that are tightly linked and, to a certain extent, addressed by using standardized technologies.
- **Cost Factor:** Installation and operation costs are important for all types of users. In the case of manufacturing companies, the installation costs of a WLAN or WPAN can be far more lower then a wired system. An additional advantage can be that the reconfiguration of their shop floors, cells cost also less than in the case of using wired networks. These aspects are valid of course for all companies.

In order to increase user productivity, combining 3G mobile networks and wireless LANs is a possible way of the future. Wireless LAN is an excellent technology to satisfy the demands of mobile employees within their enterprise locations, or in homes where they have multiple computers. Because of the very high-speed expectations and intrapremises dominated communication, the wireless LAN provides cost-effective mobility functioning as an extension to the wired LAN, optimizing private networking costs and minimizing facilities bottlenecks to speed. Outside the enterprise, 3G wireless technology provides a means for users to do what could not be done before—remain connected to their business critical applications wherever they need to go. The wide area coverage of mobile networks will make it possible for users to do business in entirely new ways, and significantly improve ease of use by using the same connection procedures wherever they may be outside of the office.

Communication Networks for Smart Organizations

Wired Communication Networks

The Internet and WWW technologies, or some of their protocols and philosophies, are applied also by the wireless communication technologies, so a short summary will be given on these technologies. Two novel, wired technologies will be also introduced, as they offer technical and service possibilities that are in certain aspects competitive with wireless technologies. PLC offers very low-cost, flexible, reliable, secure communication inside buildings or limited areas. The latest results of GRID technology make possible the very easy, integrated, and user-friendly access to databases, programs, and other computing resources. This section will be closed by the short description of VoIP, as this technology has an increasing importance in voice communication of enterprises.

Internet

Communication systems, wired as well as wireless, employ a layered approach to data transmission protocols. In the following, only the base will be introduced very briefly, in order to give a background for communication technologies presented later on.

Open systems architectures (OSA) have become an important approach to develop flexible, adaptable sets of methodologies, standards, and protocols for structured communication systems. OSA is a layered hierarchical structure, configuration, or model of a communications or distributed data processing system that enables system description, design, development, installation, operation, improvement, and maintenance to be performed at a given layer or layers in the hierarchical structure. It allows each layer to provide a set of accessible functions that can be controlled and used by the functions in the layer above it, enables each layer to be implemented without affecting the implementation of other layers, and allows the alteration of system performance by the modification of one or more layers without altering the existing equipment, procedures, and protocols at the remaining layers.

An OSA may be implemented using the open systems interconnection-reference model (OSI-RM) as a guide while designing the system to meet performance requirements. The model employs a hierarchical structure of seven layers, Each layer performs value-added service at the request of the neighboring higher layer and, in turn, requests more basic services from the next lower layer. The names of the seven layers and the protocols are shown in Table 1. In the table the security protocols are also shown; they will be discussed in the security section later on. A good and detailed work on computer networks is the book of Tanenbaum (1996).

Transmission Control Protocol/Internet Protocol - TCP/IP

The TCP/IP is two interrelated protocols that are part of the Internet protocol suite. TCP operates on the OSI Transport Layer and breaks data into packets, controlling host-to-host transmissions over packet-switched communication networks (Table 1). Internet protocol (IP) was designed for use in interconnected systems of packet-switched computer communication networks. IP operates on the OSI Network Layer and routes packets. The Internet protocol provides for transmitting blocks of data called datagrams from sources to destinations, where sources and destinations are hosts identified by fixed-length addresses.

The short descriptions of layers are as follows:

- Physical layer transmits information carrying signals over a communications channel. The communications channel (or medium) may be wireless, such as analogue or digital radio, or wired, such as optical fiber or coaxial cable.
- Link layer runs over, and utilizes, the physical layer to carry data from a transmitter to a receiver. The link layer uses frames (e.g., a collection of bytes) as its data communication unit, and is responsible for their transmission between two or more communication nodes sharing a medium. An example of a link layer standard is the (IEEE) 802.11 Wireless Local Area Network (WLAN).
- Network layer enables communication between nodes that do not necessarily share a communications medium. For packet-switched networks, the link layer carries network layer datagram units (or packets) between neighboring nodes in link layer frames. This process may require fragmentation of large packets into smaller datagrams with additional network and link layer-specific headers.
- Transport layer runs over the network layer. It may or may not ensure reliable, in-order delivery of the data generated by the application layer.
- Application layer contains functions for particular application services, such as file transfer, remote file access and virtual terminals.

Table 1. TCP/IP- and security protocols in the network

Layer Number	Layers of the OSI reference model	TCP/IP Protocols	SECURITY PROTOCOLS	
7.	Application	FTP, SMTP, TELNET, SNMP, NFS, Xwindows, NNTP, IRC, HTTP, WAP	S-HTTP, SET	S/MIME, PEM, PGP, MOSS
				SMTP
6.	Presentation	ASCII, EBCDIC, ASN1, XDR	SSL, SSH, SSH2	
5.	Session	RPC		
4.	Transport	TCP, UDP	TLS (Transport Layer Security Protocol), WAP/WTLS	
3.	Network	IP	IPv6, VPN	
2.	Data link	X.25, SLIP, PPP, Frame Relay	Electromagnetic Emission standard (89/336/EEC - European Economical Community guideline)	
1.	Physical	LAN, ARPANET		

The World Wide Web

The World Wide Web ("WWW," or simply "Web") is an information space in which the items of interest, referred to as resources, are identified by global identifiers called Uniform Resource Identifiers (URI). The term is often mistakenly used as a synonym for the Internet, but the Web is actually a service that operates *over* the Internet.

The Web is made up of three standards: The *Uniform Resource Locator* (URL), which specifies how each page of information is given a unique "address" at which it can be found; the *Hyper Text Transfer Protocol* (HTTP), which specifies how the browser and server send the information to each other; and the *Hyper Text Markup Language* (HTML), a method of encoding the information so it can be displayed on a variety of devices.

Hypertext does not have a linear order from beginning to end. It is not broken down into the hierarchy of chapters, sections, subsections, and so forth. From the Web many kinds of information are available; private persons, organizations, universities and enterprises have "home pages" that contain basic information on the owner with very easy access. Today a home page is a must-have for enterprises that want to take part in the global competition.

Grid Computing

"Grid" computing is an important new field that has to be distinguished from conventional distributed computing by its focus on large-scale resource sharing, innovative applications, and high-performance orientation. "Grid" can be defined as a hardware and software infrastructure that provides dependable, consistent, pervasive, and inexpensive access to high-end computational capabilities resulting in flexible, secure, coordinated resource-sharing among dynamic collections of individuals, institutions, and resources—to sum up, virtual organizations (Foster, 2000).

The real and specific problem that underlies the Grid concept is coordinated resource-sharing and problem-solving in dynamic, multi-institutional virtual organizations. The sharing is not primarily file exchange, but rather direct access to computers, software, data, and other resources, as is required by a range of collaborative problem-solving and resource brokering strategies emerging in industry, science, and engineering. This sharing is highly controlled, clearly and carefully defined by what is shared, who is allowed to share, and the conditions under which sharing occurs. A set of individuals and/or institutions defined by such sharing rules form is called a virtual organization (VO).

Furthermore, sharing is about more than simply document exchange (as in "virtual enterprises"): it can involve direct access to remote software, computers, data, sensors, and other resources. For example, members of a consortium may provide access to specialized software and data, and/or pool their computational resources. The members of a Virtual Organization do not necessarily have to work together on the same site, but the Grid will make it feel, to the members, as if they are on the same network.

The Grid architecture is a protocol architecture, with protocols defining the basic mechanisms by which VO users and resources negotiate, establish, manage, and exploit sharing relationships. A standards-based open architecture facilitates extensibility, interoperability, portability, and code sharing; standard protocols make it easy to define standard services that provide enhanced capabilities

The primary goal of the Grid at the moment is to allow coordinated resource sharing in virtual organizations. Current Internet technologies address communication and information exchange among computers but do not provide integrated approaches to the coordinated use of resources at multiple sites for computation. Business-to-business exchanges focus on information sharing (often via centralized servers); virtual enterprise technologies do the same. Enterprise distributed computing technologies like CORBA and Enterprise Java are not able to resource-share within the organization.

Grid is building on these existing technologies, rather than compete with them; the Grid will act as a middleware between high-level behaviors of the Internet (such as its protocols, and the lower levels, for example the application layer), complement its functionality, and to add flexibility. The Grid can be viewed as an extension to the Web, building on its protocols, and offering new functionality (Foster, Kesselman, & Tuecke, 2000).

As the Grid is built on the existing Internet, it will share its capabilities, such as simple data retrieval and transfer, as well as the basic file-sharing functions provided by peer-to peer applications. The prospects for the future, however, are far greater, and could not only change the way of sharing information, but also the way computers interpret information and even, by integrating developing technologies such as Jini and Bluetooth, how this technology can involve the daily life.

Powerline Communications

As cable, telephone, and wireless companies compete to provide high-speed Internet access to homes, a new challenger is emerging based on a decidedly old technology. The idea is to connect the Internet and network computers in a LAN, by using the world's largest existing network, the power grid. Powerline communications (PLC)—communications over the electricity distribution grid—has become a hot topic recently. Although this technology has been in use for special applications for several decades—for example, street lighting is frequently operated according to this principle—communication in these cases is exclusively in the narrowband range and transmission rates are correspondingly low.

The first attempts to realize the power grid as a communication network were not really successful, but the technological advancements over the last few years have overcome the technical issues, most notably that of line noise or interference from electrical devices plugged into the same electricity grid, which can disrupt data-transmission. PLC works by transmitting data signals through the same power cables that transmit electricity, but it uses a different frequency. To do this, every PC needs to be attached with a PLC adapter, which also functions as a modem (Hübscher, 2004).

The operation procedure of PLC can be divided into two phases:

- **Procedures, which are performed outside the home (outdoor):** The conventional telecommunications infrastructure is used to connect the relevant local network station with the telephone network or a specific Internet backbone. Depending on distance and local conditions, the connection is enabled by radio, copper lines, or optical cables. The local network station combines data and voice signals on the power grid and sends them as a data stream to any socket in connected households; that is, to the end user via the low-voltage network.
- **Procedures inside the home (indoor):** The access point forwards incoming data streams to the indoor network, and an indoor master in the household controls and coordinates all (externally and internally) transmitted data signals. Intermediate adapters separate data and power at the socket and forward the data to individual applications. There is no need for separate telephone or data cabling since the socket, far from being a mere electrical point, becomes a powerful communications interface

which bridges the last mile for high-speed Internet access, thus enabling networking throughout the building or household.

The powerline technology applied today transmits data at 4.5 Megabits per second (Mbit/s) via the electricity supply grid—in the medium-term rates of up to 20 Mbit/s are possible—and provides permanent high speed access to the Internet (always online) from every main voltage supply socket in a building, and makes broadband capacity cost-efficiently available over the "last mile." It is no longer necessary to always dial into the network, or indeed to install additional cabling within a building, so PLC is also an interesting alternative for an in-house data network.

Because PLC uses the existing electrical wirings hidden in the walls of homes and buildings, users can do away with messy cables and do not need to open floorboards, hack walls, and break ceilings to run the wires. PLC also enables indoor networking for PCs and printers, plus shared Internet access between PCs in an office or home. In addition, PLC boasts a superior distance of 300m (without using repeaters), compared to 100m for standard Fast-Ethernet and about 100m for 802.11b wireless connections.

For utility suppliers, PLC opens a whole new revenue stream for them, which they can deploy quickly. For service providers buying wholesale services from utility companies, PLC also offer various benefits, including the speed and cost of deployment and the ability to break the telephone company monopoly on last-mile access in many countries.

Voice Over Internet Protocol

Voice over Internet Protocol (VoIP) is a technology that allows people to make telephone calls using a broadband Internet connection instead of a regular (or analog) phone line. VoIP technologies convert digitized voice into data packets that are encapsulated in Internet protocol. Internet telephony refers to communications services—voice, facsimile, and/or voice-messaging applications—that are transported via the Internet, rather than the public switched telephone network (PSTN).

The basic steps involved in originating an Internet telephone call are conversion of the analog voice signal to digital format and compression/translation of the signal into Internet protocol (IP) packets for transmission over the Internet; the process is reversed at the receiving end.

There are different services included into VoIP technology. Some services using VoIP may only allow the user to call other users using the same service, but others may allow to call anyone who has a telephone number—including local, long distance, mobile, and international numbers. There are services only working over computer or a special VoIP phone, other services allow to use a traditional phone through an adaptor.

As VoIP technology unites the telephony and data worlds, companies can be the winners of this new technology because VoIP allows users to integrate their phone calls, faxes, and voice traffic over corporate Intranets and the Internet. Companies that use VoIP technologies can save a lot on long distance calls.

Making Voice over IP function efficiently in a corporate enterprise network requires adequate bandwidth allocation and management. For each call to be sent across an IP network, 17Kbps is needed of the total bandwidth. If properly designed and operated, a company's network can use a 56 or 64 Kbps link to simultaneously share several voice calls and data traffic without any delays or problems. In larger organizations, where a large amount of data is carried across a network, Voice over IP would need a separate infrastructure in order to be utilized. Especially in companies where up to 50 phone lines can be used simultaneously, an Intranet type of infrastructure will be needed to process the calls with PSPN quality.

Voice over IP technology has several advantages that will result a fast growth, especially among companies:

- Cost reduction. It is cheaper to make an IP telephony call than a circuit call because the operators avoid paying interconnect charges.
- Better utilization of infrastructure. Circuit switched telephony call takes up 64 KBPS while an IP telephony call takes up 6-8 KBPS.
- Possibilities for new added value services. Such value added opportunities include: IP multicast conferencing and telephony distance learning applications, phone directories and screen popping via IP, and "voice web browsing" where the caller can interact with a Web page by speaking commands.
- Possibility to manage a single network handling both voice and data.

IP telephony will also create great demand for new services. It will allow people to control different media and different types of terminals, such as PC and fixed phone, straight from their Web browser.

The main problems of VoIP technology applications are the interoperability, security, and bandwidth management. When these problems are solved even partially the spread of VoIP will grow extremely fast first of all among enterprises. VoIP technology can revolutionize both telecommunication services and industry and can create many advantages for companies and private users as well.

Wireless Communication Networks

Types of Wireless Networks

Wireless networks (WN) serve as the transport mechanism between and among devices and the traditional wired networks (enterprise networks and the Internet). Wireless networks have many types and are diverse, but can be categorized into four groups based on their coverage range:

- Satellite communication (SC),
- Wireless Wide Area Networks (WWAN),
- Wireless Metropolitan Area Network (WMAN),
- Wireless Local Area Networks (WLAN), and
- Wireless Personal Area (or Pico) Network (WPAN).

The main characteristics of wireless networks are as follows:

- access for anybody, from anywhere, at any time—mobility,
- on-line/real-time access,
- relative high communication speed,
- shared access to files, data/knowledge bases, and
- exchange of picture, voice—multimedia applications.

In the followings the main types of wireless networks will be introduced. In Table 2 the main characteristics of these wireless technologies are summarized.

Table 2. Main characteristics of wireless networks

Wireless network type	Operation frequency	Data rate	Operation range	Characteristics
Satellite	2170–2200 MHz	Different (9.6 kbps - 2 Mbps)	Satellite coverage	Relative high cost, availability
WWAN				
GSM (2-2.5 G)	824-1880 MHz	9.6 - 384 kbps (EDGE)	Cellular coverage	Reach, quality, low cost
3G/UMTS	1755-2200 MHz	2.4 Mbps	Cellular coverage	Speed, big attachments
iMode (3G/ FOMA)	800 MHz	64 - 384kpbs (W-CDMA)	Cellular coverage	Always on, easy to use
FLASH-OFDM	450 MHZ	Max. 3 Mbps	Cellular coverage	High speed, respond time less then 50 milliseconds
WMAN				
IEEE 802.16	2-11 GHz	Max.70 Mbps	3-10 (max. 45) km	Speed, high operation range
WWLAN				
IEEE 802.11A	5 GHz	54 Mbps	30m	Speed, limited range
IEEE 802.11b	2.4 GHz	11 Mbps	100 m	Medium data rate
IEEE 802.11g	2.4 GHz	54 Mbps	100-150m	Speed, flexibility
WPAN				
BLUETOOTH	2.4 GHz	720 kbps	10 m	Cost, convenience
UWB	1.5 – 4 GHz	50-100 Mbps	100-150 m	Low cost, low power
ZigBee	2.4 GHz, 915 - 868 Mhz	250 Kbps	1-75 m	Reliable, low power, cost effective
Infrared	300 GHz	9.6 kbps-4Mbps	0.2-2 m	Non interfere, low cost
RFID	30-500 KHz 850-950 MHz 2.4-2.5 GHz	linked to band-width, max. 2 Mbps	0.02–30 m	High reading speeds, responding in less than 100 milliseconds

Satellite Communication

Satellite communication systems offer the users independent and reliable communication with any chosen subscriber in the world in the covered area, offering high quality telephone calls, fax transmissions, high-speed data access, and e-mail messaging. Today satellite communication providers purchase services and phone sets that are appropriate for dual mode, such as Satellite-GSM Phones.

These handheld satellite telephones provide voice, fax, Internet access, short messaging, and remote location determination services (GPS) in the covered area. All of this is provided through geosynchronous satellites, but when satellite coverage is not necessary, the handset can also access the GSM cellular network. Fax and digital data is transmitted at 9600 bps throughputs, but in case users need high-speed Internet access (144 kbps) this also can be achieved by using special lightweight terminals. In special cases, 2 Mbps data transmission rate can be achieved (Sheriff & Fun Hu, 2001).

A satellite phone can fulfill all the requirements regarding mobile communications in many application fields. Interdisciplinary applications are paralleled with equipment functionality—currently available satellite phones have the size of a standard notebook computer. Their use does not require complicated procedures; activation and call charging is done similarly to cellular phone networks (Elbert, 2004).

Wireless Wide Area Networks (WWAN)

Mobile Phone

Mobile communication is connected to using mobile phones. Mobile phone was the device that offered for a great number of people the possibility to make contact with others from anywhere, at anytime, and for anybody. Mobile phone is the device that realizes mobility on a society level, as in many countries more than 70% of the population has a mobile phone.

There are different mobile systems/network protocols, which are developing very fast.

- **CDMA (Code Division Multiple Access—2G):** CDMA networks incorporate spread-spectrum technology to gracefully allocate data over available cells.
- **CDPD (Cellular Digital Packet Data—2G) :** CDPD is a protocol built exclusively for sending wireless data over cellular networks. CDPD is built on TCP/IP standards.
- **GSM (Global System for Mobile Communications—2G):** GSM networks, mainly popular in Europe.
- **GPRS (General Packet Radio Service—2.5 G) :** GPRS technology offers significant speed improvements over existing 2G technology.
- **iMode (from DoCoMo—2.5G):** iMode was developed by DoCoMo and is the standard wireless data service for Japan. iMode is known for its custom markup language enabling multimedia applications to run on phones.

- **3G:** 3G networks promise speeds rivaling wired connections. Both in Europe and North America, carriers have aggressively bid for a 3G spectrum but no standard has yet emerged.

The introduction of WAP (Wireless Application Protocol) was a big step forward for the mobile communication as this protocol made it possible to connect mobile devices to the Internet. By enabling WAP applications, a full range of wireless devices, including mobile phones, smart-phones, PDAs, and handheld PCs, gain a common method for accessing Internet information. The spread of WAP became even more intensive as the mobile phone industry actively supported WAP by installing it into the new devices. WAP applications exist today to view a variety of WEB content, manage e-mail from the handset and gain better access to network operators' enhanced services. Beyond these information services, content providers have developed different mobile solutions, such as mobile e-commerce (mCommerce).

Mobile technology affects the operation of enterprises as well. The main reasons to develop a mobile solution in the enterprise are listed in the following:

- Provide access to company e-mail,
- Provide access to Intranet applications,
- Develop specific company applications,
- Permanent contact with service workers,
- Improve work scheduling,
- Possibility for mCommerce.

Mobile communication extends company data, back-end information systems, and e-mail to mobile employees and broadens the accessibility of mission critical data. Mobile access modifies the way workers interact with colleagues, customers, and suppliers.

FLASH-OFDM

FLASH-OFDM (Fast, Low-latency Access with seamless Handoff—Orthogonal Frequency Division Multiplexing) is a cellular, IP-based broadband technology for data services on the 450 MHz band. It has full cellular mobility,

3.2 Mbps peak data rates, 384 kbps at the edge of the cell and less than 20 milliseconds of latency.

The FLASH-OFDM system consists of an airlink, an integrated physical and media access control layer, and IP-based layers above the network layer (layer 3). The IP-based layers support applications using standard IP protocols.

Radio-router technology uses a radio-transmission framework for packet-based, broadband, IP wireless communications. Radio-router technology is designed to make links in an IP network mobile. A radio-router network can be built atop the existing IP infrastructure, and since IP network technology is already well-developed and inexpensive, radio-router systems will be relatively easy, quick, and economical to implement (Flarion, 2004).

FLASH-OFDM is a wide-area technology enabling full mobility up to speeds of up to 250 km/h (critical to vehicle and rail commuters). Its ability to support a large number of users over a large area, and nationwide build outs (via wireless carriers), will do for data as the cellular networks did for voice. The IP (Internet Protocol) Interfaces In Flash-OFDM Enable operators to offer their enterprise customers access to their LANs (Local Area Networks) and users the benefits of the mobile Internet.

FLASH-OFDM support voice-packet-switched voice, not circuit-switched voice, Radio routers, IP routers with radio adjuncts, would handle packet traffic, and serve as the equivalent of cellular base stations. Consumers would connect with Flash-OFDM networks via PC cards in their notebooks and via flash-memory cards in handheld devices.

Wireless Metropolitan Area Network (WMAN)

The term WiMAX (Worldwide Interoperability for Microwave Access) has become synonymous with the IEEE 802.16 Metropolitan Area Network (MAN) air interface standard. Metropolitan area networks or MANs are large computer networks usually spanning a campus or a city. They typically use optical fiber connections to link their sites.

WiMAX is the new shorthand term for IEEE Standard 802.16, also known as "Air Interface" for Fixed Broadband Wireless Access Systems. In its original release (in early 2002) the 802.16 standard addressed applications in licensed bands in the 10 to 66 GHz frequency range and requires line-of-sight towers called fixed wireless. Here a backbone of base stations is connected to a public network, and each base station supports hundreds of fixed subscriber stations, which can be both public Wi-Fi "hot spots" and enterprise networks with firewall.

The 802.16a extension, (issued in January 2003), covers non-line of sight (NLOS) applications and allows use of lower frequencies (2 to 11 GHz), many of which are unregulated. The 802.16a extension is a wireless metropolitan area network (WMAN) technology that will connect 802.11 hot spots to the Internet and provide a wireless extension to cable and DSL for last mile broadband access. The 802.16a provides up to 50 km of linear service area range. The technology also provides shared data rates up to 70 Mbps, which is enough bandwidth to simultaneously support more than 60 businesses with T1-type connectivity and hundreds of homes at DSL-type connectivity.

The 802.16e version is under development (scheduled publication is October 2005), and is expected to support mobile wireless technology—that is, wireless transmissions directly to mobile end users, so mobile applications thus enable broadband access directly to WiMAX-enabled portable devices ranging from smartphones and PDAs to notebook and laptop computers. This will be similar in function to the General Packet Radio Service and the "one times" radio transmission technology (1xRTT) offered by phone companies.

WiMAX is considered the next step beyond Wi-Fi because it is optimized for broadband operation, fixed and later mobile, in the wide area network. It already includes numerous advances that are slated for introduction into the 802.11 standard, such as quality of service, enhanced security, higher data rates, and mesh and smart antenna technology, allowing better utilization of the spectrum.

Wireless Local Area Network (WLAN)

Local area wireless networking, generally called Wi-Fi (also known as 802.11b Ethernet) is a hot topic. Companies, universities, and home users are setting up wireless access points and running notebook computers without network wires.

Wi-Fi, or Wireless Fidelity, allows users to connect to the Internet from their home, from a hotel room, or from a conference room at work without wires. Wi-Fi enabled computers send and receive data anywhere within the range of a base station with a speed that is several times faster than the fastest cable modem connection.

Wi-Fi connects the user to others and to the Internet without the restriction of wires, cables or fixed connections. Wi-Fi gives the user freedom to change locations (mobility)—and to have full access to files, office, and network connections wherever she or he is. In addition Wi-Fi will easily extend an established wired network (Anderson, 2003).

Wi-Fi networks use radio technologies called IEEE 802.11b or 802.11a standards to provide secure, reliable, and fast wireless connectivity. A Wi-Fi network can be used to connect computers to each other, to the Internet, and to wired networks (which use IEEE 802.3 or Ethernet). Wi-Fi networks operate in the 2.4 (802.11b) and 5 GHz (802.11a) radio bands, with an 11 Mbps (802.11b) or 54 Mbps (802.11a) data rate or with products that contain both bands (dual band), so they can provide real-world performance similar to the basic 10BaseT wired Ethernet networks used in many offices. 802.11b has a range of approximately 100 meters.

Products based on the 802.11a standard were first introduced in late 2001. Its strengths are the high speed and lower risk of radio frequency interference than either 802.11b or 802.11g. Its weakness is that "*a*" is incompatible with the more popular "*b*" and the emerging "*g,*" because it strayed from the 2.4-GHz band. As WLAN is spreading, it could prove essential to serving large populations in a concentrated area, such as downtowns, universities, and business centers.

The 802.11g promises complete interoperability with "b" and transmission rates up to five times faster in the same 2.4-GHz band. Early products are already on the market. The higher vulnerability to radio frequency interference from other 2.4-GHz devices (late-generation cordless phones) is a big challenge for 802.11g (Engst & Fleishman, 2003).

Wi-Fi networks can work well both for home (connecting a family's computers together to share such hardware and software resources as printers and the Internet) and for small businesses (providing connectivity between mobile salespeople, floor staff, and "behind-the-scenes" departments). Because small businesses are dynamic, the built-in flexibility of a Wi-Fi network makes it easy and affordable for them to change and grow.

Large companies and universities use enterprise-level Wi-Fi technology to extend standard wired Ethernet networks to public areas like meeting rooms, training classrooms, and large auditoriums, and also to connect buildings. Many corporations also provide wireless networks to their off-site and telecommuting workers to use at home or in remote offices.

It is easy to extend the existing networks with a Wi-Fi LAN to add another wireless computer to a Wi-Fi network. There is no need to purchase or lay more cable or find an available Ethernet port on the hub or router; just the card has to be plugged in to the computer and it is connected to the network.

Wireless Personal Area (or Pico) Network (WPAN)

WPAN represents wireless personal area network technologies such as Ultra-wideband (UWB) (Kelland, 2003), ZigBee, Bluetooth, RFID, WiMedia and IrDA.

Designed for data and voice transmission, low data rate standards include ZigBee, (IEEE 802.15.4) (Karayannis, 2003) and Bluetooth (IEEE 802.15.1), and enables wireless personal area networks to communicate over short distances, generating a new way of interacting with our personal and business environment. ZigBee provides ultra-low cost solutions for applications requiring low data rates and long battery-life, such as remote controls and sensor applications (free frequency bands including 2.4 GHz, 915 MHz and 868 MHz, transmission ranges of 30-100 m are possible). In the home, this will mean a single remote control device will operate TVs, DVD players, audio systems, and other entertainment and computing equipment, as well as controlling lights, heating, home appliances, security systems, and even toys. Bluetooth chipsets, now on their third and fourth generation, are targeted primarily at the cell phone and PC peripheral industries (2.4 GHz band, peak data throughput of 720 KBps, distances about 10 m).

Two technologies of the WPAN group the RFID and Bluetooth will be introduced more detailed in the following.

Radio Frequency Identification (RFID)

The main purpose of the RFID (radio frequency identification) technology is the automated identification of objects with electromagnetic fields. An RFID system has three basic components: transponders (tags), readers (scanners), and application systems for further processing of the acquired data. There is a large variety of different RFID systems; they can use low, high, or ultra-high frequencies, tags may emanate only a fixed identifier or can have significant memory and processing capabilities. Transponders can contain effective security protocols or no security features at all. Most of the tags have passive powered by the radio field emitted by the reader but there are also active tags with a separate power supply (Finkenzeller, 2003).

RFID systems can be distinguished according to their frequency ranges. Low-frequency (30 KHz to 500 KHz) systems have short reading ranges and lower system costs. They are usually used in, for example, security access and animal identification applications. High-frequency (850 MHz to 950 MHz and 2.4 GHz to 2.5 GHz) systems, offering long read ranges (greater than 25 meters) and high reading speeds, are used for such applications as railroad car tracking and automated toll collection. However, the higher performance of high-frequency RFID systems generate higher system costs.

In determining data transfer rates, carrier wave frequency is of primary importance. In general, a higher data transfer belongs to a higher frequency. In case of the 2.4-2.5 GHz spread spectrum band, 2 megabits per second data rates can be achieved. Spread the spectrum apart, and increasing the bandwidth allows an increase in noise level and a reduction in signal-to-noise ratio, so bandwidth is an important consideration in this respect.

According to the reader's power output and the used radio frequency, the range extends from a few centimeters to 30 meters or more. When an RFID tag passes through the electromagnetic zone, it detects the reader's activation signal. The reader decodes the data encoded in the tag's integrated circuit (silicon chip) and the data is passed to the host computer for processing.

RFID as an automatic identification technology has got an increasing importance in recent years and is now being seen as a radical means of enhancing data handling processes. A range of devices and associated systems are available to satisfy an even broader range of applications. Data within a tag may provide identification for goods in transit, an item in manufacturing process, or the identity of a vehicle, animal, or individual. By including additional data the prospect is provided for supporting applications through item-specific information or instructions immediately available on reading the tag.

Bluetooth

Bluetooth is a short-range radio device that replaces cables with low-power radio waves to connect electronic devices, whether they are portable or fixed. It is a wireless personal area network (WPAN) specified in IEEE 802.15, Working Group for wireless personal area networks. Bluetooth, named after Harald Bluetooth, the 10th century Viking king, is a consortium of companies (3Com, Ericsson, Intel, IBM, Lucent Technologies, Motorola, Nokia, and

Toshiba) bonded together to form a wireless standard (Bakker & Gilster, 2002).

The Bluetooth device also uses frequency-hopping to ensure a secure, quality link, and it uses ad hoc networks, meaning that it connects peer-to-peer. When devices are communicating with each other they are known as piconets, and each device is designated as a master unit or slave unit, usually depending on who initiates the connection. However, both devices have the potential to be either a master or a slave.

The Bluetooth user has the choice of point-to-point or point-to-multipoint links, whereby communication can be held between two devices, or up to eight.

Bluetooth is not only a standard, but it is also a product. The hardware consists of a microchip with a radio transceiver. It can be incorporated into a laptop or wireless phone. It can access other ad hoc networks or local access points. It is a short-range system, operating at a normal range of 10m (0 dBm) and an optional range of 100 m (+20 dBm). It uses 2.4 GHz as its base frequency. It can reach 6 Mbps in a multiple piconet ad hoc structure.

The new version of Bluetooth, Bluetooth 2.0+EDR (Enhanced Data Rate) offers a significant speed over version 1.2. While Bluetooth 1.2 offers a data transfer rate of 1 Mbps, Bluetooth 2.0 offers data transfer speed 3Mbps. As an addition, a significant increase in power efficiency results in longer battery life—that is a very important factor for mobile users. The new version is backwards-compatible to earlier versions and allows the connection of multiple devices.

Trust and Security in Smart Organizations

Role of Trust in Smart Organizations

Trust and Trust Building in SO

In all information and communication systems there is a common factor: the human being. This factor plays the most important role in every level and in every aspect. A human can be a designer, a developer, or a user of the system. The most frequent instantiation of the human being is the average user, who maybe is not well-informed or skilled in computer science, but has his or her own personality and psyche.

In order to move the individuals to use a certain information system, they have to be convinced that it is safe to use the system, that their data will not be modified, lost, used in other way as defined previously, and so forth. If the individuals have been convinced they will trust the system and they will use it.

Trust can be defined as a psychological condition comprising the trustor's intention to accept vulnerability based upon positive expectations of the trustee's intentions or behavior (Rousseau, Sitkin, Burt, & Camerer, 1998). Those positive expectations are based upon the trustor's cognitive and affective evaluations of the trustee and the system and world, as well as of the disposition of the trustor to trust. Trust is interpreted in terms of expectation, attitude, willingness, or perceived probability. Trust can cause or result from trusting behavior (e.g., cooperation, taking a risk) but is not behavior itself.

The following components are included into most definitions of trust:

- willingness to be vulnerable or to rely,
- confident, positive expectation and positive attitude towards others,
- risk and interdependence as necessary conditions.

Trust has different forms, such as

- **Intrapersonal trust:** trust in one's own abilities; self-confidence basic trust (in others).
- **Interpersonal trust:** expectation based on cognitive and affective evaluation of the partners; in primary relationships (e.g., family) and non-primary relationships (e.g., business partners).
- **System trust:** trust in depersonalized systems or the world that function independently (e.g., economic system, regulations, legal system, technology); requires voluntary abandonment of control and knowledge (Luhmann, 1979).
- **Object trust:** trust in non-social objects; trust in its correct functioning (e.g., in an electronic device).

Collaboration and cooperation are main characteristics of networked organizations, so the contacts among the users, the human beings, have outstanding importance. A very important element of this human contact is trust. In a networked organization, trust is the atmosphere, the medium in which actors are moving (Castelfranchi & Yao-Hua, 2001). Trust is the base of cooperation, the normal behavior of the human being in the society. The ability of enterprises to form networked systems depends on the existing level of trust in the society and on the capital of society (Fukuyama, 1995). As the rate of cooperation is increasing in all fields of life, the importance of trust is evolving even faster.

In this new organizational environment, new methods and techniques of trust building have to be developed, as the conventional rules cannot be applied. According to different experiments, the level of trust between cooperating virtual teams is highly influenced by the type of communication and the duration of contacts.

Technology Component of Trust

In building trust there are two approaches: information technology approach, and human centered approach, based on culture and morality. Information technology approach means that security has to increase by different architectures, protocols, certifications, cryptography, authentication procedures and standards, and this increased security generates the trust of users. This means access control (passwords, firewalls), protect integrity and privacy of messages and databases (cryptography), and identification of users. Parallel stressing the effectiveness of these technologies for the humans (users) can cause that they will trust in the systems based on these convincing actions. Based on the technological approach, 100% security never can be obtained (there will be always security holes somewhere in the systems), so full trust can not guaranteed based on these mechanisms.

Human Side of Trust Building: Feeling of Trust

The feeling of security experienced by a user of an interactive system does not depend on technical security measures alone. Other (psychological) factors can play a determining role; the user's feeling of control can be one of these factors.

It has to be analyzed why people feel safe and secure, what causes these feelings. The hypothesis of D'Hertefelt (2000) was that "The feeling of security experienced by a user of an interactive system is determined by the user's feeling of control of the interactive system." The more a user feels in control of an interactive program, the more the user will trust the site and the program. From this aspect user interface has the main role, that is, the menu structure, the messages send for the user by the system. In the case the user feels that it is easy to use, it is transparent; he or she can control the system (even with low-level computer knowledge)—that is, the system is "user friendly," and through this he can be convinced that he is using a trustworthy system.

It would be a mistake to think that applying only the human-centered approach would generate trust; the technological part has to be added as well (e.g., biometrical identification), so the structured integration of the two approaches can result in the expected level of trust.

Relation of Trust and Time

Trust is a dynamic process and it alters based on experience. The trusting process begins when an individual perceives indications that suggest a person or organization may be worthy of trust. These indications can include behaviors such as manners, professionalism, and sensitivity, and these forms are designed to represent trustworthiness. These formal claims to trustworthiness become strengthened over time and are eventually transformed into "character traits," such as dependability, reliability, and honesty.

The process of building trust is slow; trust is formed gradually, and it takes quite a lot of time and repeated positive experiences (Cheskin, 1999). Online trust can be described as a kind of human relationship. The initial stage is that of interest and distrust; there has to be a motivation, a need, to get interested in the service, or coworking. In subsequent phases the trust will evolve, or in the case of negative experiences the cooperation will terminate.

Trust depends on the time span of cooperation and the type of connection as well. It can be stated that there are differences in the trust-building process in short-term and long-term relationships. In the case of short-term relationships (e.g., in a virtual organization), trust must be achieved quickly, and then maintain with no, or rare face-to-face interaction. The members of these teams must assume that other remote team members are trustworthy, and then later on modify their assumptions according their positive or negative experiences.

In long-term relationships there are four factors that influence trust-building (Rocco, Finholt, Hofer, & Herbsleb, 2001):

- expectation of future interaction may motivate greater investment in building trustworthy relationships,
- long-term relationships offer more time to establish trustworthiness through routines and culture,
- people have more communication channels, which may affect trust to the extent that participants have additional ways to clarify misunderstandings or to correct failures,
- participants are interested in successful task performance, and trust formation may assume a higher priority.

Developing Trust in Networked Organizations

Today the different types of networked organizations need new types of cooperation; as the members of the working teams are geographically (physically) separated, they use shared documents, and communicate through e-mail and high-quality audio and video channels. These teams are called "virtual teams," as they never meet personally and they have no face-to-face (FTF) contact. The work of teams without FTF contact is less effective and reliable based on the observation stated by Handy (1995) that "trust needs touch." According to case studies, it is evident that trust of virtual team members is significantly lower than trust in conventional teams (Rocco, Finholt, Hofer, & Herbsleb, 2001). In other experiments where interaction was primarily via e-mail, very similar results have gained as in geographically distributed teams (Jarvenpaa & Leidner, 1999)

In an experiment introduced in Bos (2002), four media types were compared: Chat (text), phone conference, videoconference, and face-to- face. Chat was significantly worse than each of the other three conditions, but audio and video did as well as face-to-face in overall cooperation, and were a definite improvement over text-chat only CMC. However, these two channels still showed evidence of delayed trust, in that they took longer to reach high levels of cooperation.

The latest research shows that if people meet before using computer-mediated communication, they trust each other, as trust is being established through touch. In case participants do not meet formally but they initiate various getting-acquainted activities over a network, trust is much higher than if they do nothing before, and nearly as good as a prior meeting. Using chat forums to get acquainted is nearly as good as meeting, and "even just seeing a picture is better than nothing" (Zheng, Veinott, Bos, Olson, Gary, & Olson, 2002).

Security in Smart Organizations

What is Security?

Security is a conscious risk-taking, so in every phase of a computer system's life cycle must be applied that security level which costs less than the expense of a successful attack. In other words, security must be so strong that it would not be worth it to attack the system, because the investment of an attack would be higher than the expected benefits. Data and system security are needed for all types of network-connected devices, such as a personal computer or a mobile phone. Data security involves protecting user and administrative data against unauthorized monitoring or modification, while system security deals with the consistency of the system component itself. As an addition, communication systems must address the issue of network security, which involves vice versa protection of the user device and the network. In Table 3, the main practical fields of ICT security are summarized.

In essence, security is a practice of risk management. There is no open system that is completely secure, and increases in system security typically decrease

Table 3. Main fields of ICT security

	Organization security	Personal security	Network (channel) security	Computer (end point) security
Human & SW security	Security policy (e.g., access rights).	Trained and reliable staff under control.	Using reliable network tools, and frequently checked communication channels and well configured network elements.	Using tested application SW tools, and frequently checked operation system, and properly configured HW systems.
Physical security	Computers in secure places of the building and offices.	Physical identification technologies (fingerprints, etc.)	Prevent direct, or close access to network cables, or application of special technologies.	Prevent direct physical access to computers by unauthorized persons, or block close access in electromagnetic way.

system performance and usefulness. At different system levels, different security solutions have to be applied, and these separate parts have to cover the entire system consistently.

A carefully designed security system process has the following basic steps:

- Definition of threats and their attack types from which the system has to be protected.
- The degree of protection that should be applied.
- The place and mode of the protection that should be applied.
- Selection of security mechanisms and services.
- Selection of HW and SW solutions.

To develop the proper security policy, to select the proper equipment, tools, and the best-fitting methodology, algorithm needs high-level expertise as in case such a multidimensional, interdisciplinary decision problem there is no optimal, only suboptimal solution in many cases.

The problem space is extremely complex, as the whole economy is based on networked information management and all sectors are strongly influenced by the ICT, and in the information society the behavior and habits of the people are dynamically changing, and government supported programs can speed up certain processes.

Threats and Trends of Cyber Crimes

The first step is the definition of the threat model of the ICT system. The threat model is the collection of probable attack types, so it defines the system protection requirements as well. In the following, the active attacks will be described, but passive attacks precede active attacks in many cases.

In order to define a very general, but at the same time very typical, attack tree the CSI statistics will be used. The "Computer Crime and Security Survey" of the Computer Security Institute (CSI) is based on responses from 486 computer security practitioners in U.S. corporations, government agencies, financial institutions, medical institutions, and universities (FBI, 2004). The survey confirms that the threat from computer crime and other information security breaches continues unabated.

The total reported financial loss of 251 responders was $141,496,560 in 2004, while in 2000 this sum was $265,589,940 of 249 responders. These numbers demonstrate that the value or the loss and damage caused by the attacks is decreasing. One reason of this shrinkage could be that the companies use security technologies today to a greater extent than they did several years before. The 483 responders use the following security technologies (in %): smart cards-35, intrusion detection-68, encrypted login-56, firewalls-98, anti-virus SW-99, encrypted files-42, biometrics-11, access control-71. The most frequent types of attacks and the financial loss caused by them are listed in Table 4. (The percentage gives the rate of responders involved in the attack; the losses are in USD).

Security Services and Mechanisms

The following services form together the sense of "trust" for a human being who uses a service or a given equipment (Menezes, van Oorschot, & Vanstone, 1996):

- Confidentiality/privacy ensures that only the sender and the intended recipient of an encrypted message can read the contents of that message. To guarantee privacy, a security solution must ensure that no one can see, access or use private information, such as addresses, credit card information, and phone numbers, as it is transmitted over the Internet.
- Integrity ensures the detection of any change in the content of a message between the time it is sent and the time it is received. In many systems, if

Table 4. Most frequent types of attacks in US (FBI, 2004)

Type of attack	In %	Caused financial loss
Virus	78	55,053,900
Insider abuse of net access	59	10,601,055
Laptop/mobile theft	49	6,734,500
System penetration	39	901,400
Unauthorized access	37	4,278,205
Denial of service	17	26,064,050
Abuse of wireless networks	15	10,159,250
Thief of proprietary information	10	11,460,000
Misuse of public WEB	10	2,747,000
Web site defacement	7	958,100
Telecom fraud	5	3,997,500

an alteration is detected, the receiving system requests that the message be resent.

- Authentication ensures that all parties in a communication are who they claim to be. Server authentication provides a way for users to verify that they are really communicating with the Web site they believe they are connected to. Client authentication ensures that the user is who they claim to be.
- Non-repudiation provides a method to guarantee that a party to a transaction cannot falsely claim that they did not participate in that transaction. In the real world, hand-written signatures are used to ensure this.

The means for achieving these services depends on the collection of security mechanisms that supply security services, the correct implementation of these mechanisms, and how these mechanisms are used. Three basic building blocks of security mechanisms are used:

- Encryption is used to provide confidentiality and can provide authentication and integrity protection.
- Digital signatures are used to provide authentication, integrity protection, and non-repudiation.
- Checksums/hash algorithms are used to provide integrity protection and can provide authentication.

One or more security mechanisms are combined to provide a security service, and a typical security protocol provides one or more services. As there are too many security technologies, tools, and equipment to be introduced in this place, only the most frequently used, or some new ones, will be shortly described in the following. Detailed descriptions can be found in Menezes et al. (1996), Anderson (2001), and Schneier (1996).

Tools, Methods, and Techniques for Security

Achieving Confidentiality

The main factor of trust is confidentiality, which can be achieved by technologies that convert or hide the data or text in a form that cannot be interpreted by unauthorized persons. There are two major techniques to fulfill this goal: encryption and steganography.

- Encryption is transforming the message to a ciphertext, so that an enemy who monitors the ciphertext cannot determine the message sent. The legitimate receiver possesses a secret decryption key that allows him to reverse the encryption transformation and retrieve the message. The sender may have used the same key to encrypt the message (with symmetric encryption schemes) or used a different, but related key (with public key schemes). Public key infrastructure (PKI) technology is widely used, as DES and RSA are well-known examples of encryption schemes, while the AES (with the Rijndael algorithm) belongs to the new generation.
- Steganography is the art of hiding a secret message within a larger one in such a way that the opponent cannot discern the presence or contents of the hidden message. For example, a message might be hidden within a picture by changing the low-order pixel bits to be the message bits.

Security Architectures

The security architectures represent a structured set of security functions (and the needed hardware and software methods, technologies, tools, etc.) that can serve the security goals of the distributed system. In addition to the security and distributed enterprise functionality, the issue of security is as much (or more) a deployment and user-ergonomics issue as technology issue. That is, the problem is as much trying to find out how to integrate good security into the industrial environment so that it will be used, trusted to provide the protection that it offers, easily administered, and really useful.

Firewalls

Firewalls can make the user's network appear invisible to the Internet, and they can block unauthorized and unwanted users from accessing files and systems. Hardware and software firewall systems monitor and control the flow of data in and out of computers in wired and wireless enterprise, business, and home networks. They can be set to intercept, analyze, and stop a wide range of Internet intruders and hackers. Like VPNs, there are many types and levels of firewall technology. Many firewall solutions are software only; many are powerful hardware and software combinations.

Virus Defense

Viruses and other malicious code (worms and Trojans) can be extremely destructive to the vital information and the computing systems both for individuals and businesses systems. There are big advances in anti-virus technology, but malicious codes remain a permanent threat. The reason is that the highest-level security technology can be only as effective as the users who operate them. In the chain of computer security, human beings seem to be the weakest point, so there is no absolute security in virus defense.

Virtual Private Network (VPN)

Today most companies use VPN to protect their remote-access workers and their connections. It works by creating a secure virtual "tunnel" from the end-user's computer through the end-user's access point or gateway, through the Internet, all the way to the servers and systems of the company. It works both for wired and wireless networks and can effectively protect transmissions, for example, from Wi-Fi equipped computers to corporate servers and systems.

A VPN works through the VPN server at the company headquarters, creating an encryption scheme for data transferred to computers outside the corporate offices. The special VPN software on the remote computer or laptop uses the same encryption scheme, enabling the data to be safely transferred back and forth with no chance of interception. However, VPN access, which enables access to the company network, e-mail, and communications systems, is provided only to those who've been given authorization.

To provide a full VPN service, wireless communications systems must support the following requirements:

- broadband data access (e.g., average sustainable throughput of 1.5 Mbps downlink/600 Kbps uplink),
- low latency connectivity (~35 ms roundtrip),
- end-to-end security (authentication, integrity protection, and/or encryption,
- full mobility (at vehicular speeds).

VPN remote access protocols typically reduce throughput by 25-30%. Even without this drawback, dial-up and 2.5G data services (at around 40-56 Kbps) do not support a satisfactory remote VPN experience. VPN requires a broadband connection, particularly for file sharing, enterprise-specific applications, and large e-mail attachments so critical to today's business communications.

Identification of Persons

Generally, biometrics refers to the study of measurable biological characteristics. In computer security, biometric technologies are defined as automated methods of identifying or authenticating the identity of a living person based on his or her physiological (e.g., fingerprint, hand, ear, face, iris, or retina) or behavioral (e.g., signature, voice, keystroke) characteristic. This method of identification is preferred over current methods involving passwords and pin numbers, as the person to be identified is required to be physically present at the point-of-identification, so the person of user is identified, not the device, as in case of PIN and password.

Biometric techniques provide a significantly higher level of identification than passwords or smart cards alone. Because biometric characteristics are unique to each individual, they can be used to prevent theft or fraud. Unlike a password or personal identification number (PIN), a biometric trait cannot be forgotten, lost, or stolen. According to security experts, biometrics is considered as providing the highest level of security. Biometry can be used in IC systems instead of passwords, as with biometry the person can be identified, not the device.

Smart Cards

There is a strong need for a tool that can fulfill the functions connected to trustworthy services. Smart card (SC) technology can offer a solution for current problems of secure communication by fulfilling simultaneously the main demands of identification, security, and authenticity besides the functions of the actual application.

Smart card is a plastic plate that contains a microprocessor, a chip, similar to computers. It has its own operation system, memory, file system, and interfaces. A smart card can handle all authorized requests coming from the "outside world." There are different SC configurations equipped with different interfaces. The crypto-card has a built-in chip for doing encryption and decryption, other cards have keyboards, and the SC for secure identification has fingerprint sensor (Koller, 2001). Smart cards can help in secure signing of digital documents, as well. Smart cards can be read by SC-readers integrated or connected to PCs or any other equipment. Smart cards are also important parts of physical or logical access systems for enterprises.

The application of SCs in the security field can result in the next step of the technological revolution because of new possibilities in effective integration of the functions of security and the actual application field. In this way, the SC can be the general, and at the same time personalized, "key" of the citizens for the information society.

Personal Trusted Device

People like smart, small equipment, tools that they can keep in their hands, can bring them permanently with them, so they can control them both physically and in time. This physical and time controllability makes people think these devices are secure (physically nobody else can access them), so they trust them (even this approach is not always really true). In the case that such a device can be used for communication, it is called mobile phone.

Today mobile phones represent the first generation of Personal Trusted Devices (PTD) as they can be used not only for talking but for different other functions as well, and PKI and other crypto-systems can be installed on them. The user authentication could be done based on biometry (fingerprint or voice). The connection of mobile phones with the Internet made a big leap in the direction to become mobile phones to PTD. The scale of functions became really wide and different mobile technologies have appeared (*m*Technologies).

Security Standards

In the field of security standards and quasi standards have an important role. In the following some of the most relevant ones are introduced briefly, only to show the directions and status of these significant works.

In order to classify the reliability and security level of computer systems, an evaluation system has been developed and the criteria have been summarized in the so-called "Orange book" (Orange book, 1985). Its purpose is to provide technical hardware/firmware/software security criteria and associated technical evaluation methodologies in support of the overall ADP system security policy, evaluation, and approval/accreditation responsibilities promulgated by DoD Directive 5200.28.

The ISO/IEC 10181- multi-part (1-8) "International Standard on Security Frameworks for Open Systems" addresses the application of security services in an "Open Systems" environment, where the term "Open System" is taken to include areas such as database, distributed applications, open distributed processing, and OSI (ISO, 1996). The Security Frameworks are concerned with defining the means of providing protection for systems and objects within systems, and with the interactions between systems. The Security Frameworks are not concerned with the methodology for constructing systems or mechanisms. The Security Frameworks address both data elements and sequences of operations (but not protocol elements), which may be used to obtain specific security services. These security services may apply to the communicating entities of systems as well as to data exchanged between systems, and to data managed by systems.

The ISO/IEC 15408 standard (ISO, 1999) consists of three parts, under the general title "Evaluation Criteria for Information Technology Security" (Part 1: Introduction and general model, Part 2: Security functional requirements, Part 3: Security assurance requirements). This multipart standard defines criteria to be used as the basis for evaluation of security properties of IT products and systems. This standard originates from the well-known work called "Common Criteria" (CC). By establishing such a common criteria base, the results of an IT security evaluation will be meaningful to a wider audience.

Security in Networked Environments

In networked enterprises such as smart organizations the goal for security is to reflect, in a computing and communication-based working environment, the general principles that have been established in society for policy-based resource access control.

Each involved entity or node should be able to make their assertions without reference to a mediator and especially without reference to a centralized mediator (e.g., a system administrator) who must act on their behalf. Only in this way will computer-based security systems achieve the decentralization needed for scalability in large distributed environments.

The resource access control mechanisms should be able to collect all of the relevant allegations and make an unambiguous access decision without requiring entity-specific or resource-specific local, static configuration information that must be centrally administered.

In order to be the security a successful part of the distributed, networked environment—providing both protection and policy enforcement—each principal entity should have neither more nor less involvement than they do in the currently established procedure that operates in the absence of computer security. Only the form has to be changed, such as a digital signature instead of signing a paper. In case of such system, this sort of a security infrastructure should provide the basis of automated management of resources that precedes the construction of dynamically and just-in-time configured systems to support different user defined application-oriented requirements.

The expected advantage of computer-based systems is in maintaining access control policy, but with greatly increased independence from temporal and spatial factors (e.g., time zone differences and geographic separation), together with automation of redundant tasks such as credential checking and auditing. The security architectures represent a structured set of security functions (and the needed hardware and software methods, technologies, tools, etc.) that can serve the security goals of the distributed system.

As an addition to the "classic" networked environments, companies aspire to mobilize their workforces through application of wireless communication technology. Mobility brings obvious advantages to the enterprises, including increased productivity, improved communications, cost reductions, and revenue creation.

But as the amount of information that is transported over a wireless link increases, security attentions increase. The same security troubles that currently exist in the wired network, fear of identity theft and unauthorized monitoring of financial information and trade secrets, also apply to the wireless world. In addition, the wireless network is an open and easily accessible medium. To address these concerns, carriers must deploy security solutions for wireless technologies that are secure enough to satisfy stringent enterprise requirements.

Wireless communication systems (as well as wired systems) employ a "layered" network protocol design. The physical-, link-, network-, transport-, and application layers are each responsible for certain functions in the network whereby appropriate security mechanisms are applied at each layer to ensure the optimum combination of high security and peak network performance.

Network Security in Smart Organization

In this subchapter, the security characteristics of the different network types will be introduced briefly. The technologies applied in wired networks will be presented first, as many of them are applied in the wireless networks as well. There are four different concerns that all security systems can address: privacy (confidentiality), integrity, authenticity, and non-repudiation. This is the goal in the case of the different networks as well, independently what type of media they use for data transmission.

Wired Network Security

Internet and WWW

At the beginning of networking there was a need mainly for the reliable operation, but the secure and authentic communication has became a key factor for today. According to Internet users, security and privacy are the most important functions to be ensured and by increasing the security the number of Internet users could be double or triple according to different surveys. The main reason for the increased demand is the spread of electronic commerce through the Internet, where money transactions are made in the amount of millions of dollars a day. It is not just the question of content of the communication or the user account—it is the question of money.

There are several solutions to secure the network; just security is in inverse proportion to usability and the most of the security tools are patches, extra solutions and rather stand-alone techniques. There are alternatives to use secure connections, some examples from the everyday applications (Table 1).

The FTP (file transfer protocol) application is used to provide file transfer services across a wide variety of systems. Usually implemented as application-level programs, FTP uses the Telnet and TCP protocols. The server side requires a client to supply a login identifier and password before it will honor requests. The information travels in plain, and with ftp dump is possible to sniff the communication, therefore advisable to use SSH-based SCP (secure copy) for file transfer. SSH is a Secure Shell, secure access method of a remote server instead of telnet (includes secure copy service instead of FTP, and transfers securely X sessions too).

Instead of HTTP there is SHTTP (secure hypertext transport protocol), which is HTTP over SSL (secure socket layer). Instead of simply e-mail there is the PGP (pretty good privacy) signed e-mail. With these techniques it can be guaranteed that the information in e-mail, file, or on a Web page will be reached only by authorized parties.

Over the Internet, the secure socket layer (SSL) protocol, digital certificates, and either user name/password pairs or digital signatures are used together to provide all four types of security.

SSL uses public key cryptography, bulk encryption algorithms, and shared secret key exchange techniques to provide privacy over the Internet. To provide integrity, SSL uses hashing algorithms that create a small mathematical fingerprint of a message. If any part of the message is altered, it will not match its fingerprint when the message is checked at the receiving end. In this case, the sender is asked to resend the message.

The remaining issue to address is non-repudiation. As with client authentication, most Web applications today simply rely on the entry of a user name and password to provide non-repudiation.

Applications can request a digital signature from a client, which requests that the user specifically authorize a transaction. The authorization is then encrypted utilizing the user's private key from their client certificate. Not surprisingly, a digital signature is analogous to a real signature on a check and serves the same purpose. So far though, the adoption of client certificates for use by individuals on the Internet has been slow.

Different combinations of all of these security techniques are used for different applications, depending on which forms of security are important and the degree to which the solution needs to be balanced with the convenience for the user. For example, certificate-based client authentication and non-repudiation are not widely used on the Web today because most users don't want to be bothered with the administrative tasks of obtaining and safely maintaining a client certificate.

Security in the Grid

It is important to fix that the Grid can be viewed as an "extension" of the Internet, so it is rather a set of additional protocols and services that build on Internet protocols and services to support the creation and use of computation- and data-enriched environments. Any resource that belongs to the Grid also, by definition, belongs to the Internet.

As a result of the research and development efforts of the Grid community protocols, services and tools have been produced that include, for example, security solutions that support certificate management, coordination policies and services supporting secure remote access to computing and data resources and the co-allocation of multiple resources.

With respect to security aspects of the connectivity layer of the Grid, it is obvious that the complexity of the security problem makes it important that any solutions should be based on existing standards whenever possible. As with communication, many of the security standards developed within the context of the Internet protocol suite are applicable (e.g., user "log on" [authenticate,] integration with various local security solutions, user-based trust relationships).

The public key-based grid security infrastructure (GSI) protocols are used for authentication, communication protection, and authorization. GSI builds on and extends the transport layer security (TLS) protocols to address most of the issues listed above; in particular, single sign-on, delegation, integration with various local security solutions (including Kerberos), and user-based trust relationships.

The Grid will also offer a larger variety of resources; for example, remote execution of software, use of computing power, and secure access to remote networks, similar to virtual private networks (VPN).

Security Issues in PLC

From a cyber security perspective, the electric power grids are now more fragile, and margins for error are significantly less. With diminishing margins and power reserves, the probability for cascading catastrophic effects is higher.

There are opinions that hackers could shut down the Internet and the electric power grid if they wanted to, based on some theories of how networks work. The idea that certain nodes on a network are more important than others is nothing new, but that doesn't explain how the Internet gets shut down or (even more unlikely) how a "hacker" would shut down a power grid. There are theories to suggest some useful things about how certain nodes should be even more carefully protected from such attacks.

But the highly decentralized structure of the power plants—generators are not connected to the networks, which are hooked to the Internet—means that the damage hackers can cause is limited. Power plants are complex technological organizations, so to shut down a generator, one has to open circuit breakers and instruct generators to lower the "set points," the levels at which they are transmitting power. This is not something that can be done solely via a computer network. Security experts say that energy companies are becoming increasingly sophisticated with network security, and have software systems in place allowing them to monitor any suspicious activity. That's important, because while the networks controlling power grids are currently offline, the utilities will come to rely more and more on the Internet. Companies recently launched a Web-based service for its customers, which will eventually offer services including online bill payment. This is where the companies are vulnerable; hackers could break into the network and "modify" the billing system.

However, there are potential security issues because a single power line from the utility company goes to multiple homes and office buildings. This means that hackers can "listen in" on the shared bandwidth. To further protect the data, this technology includes sophisticated encryption techniques built into the hardware (into powerline network adapters) so all data packets are automatically encrypted prior to transmission over the powerline network. Only computers that know the password can decrypt the packets and read data.

VoIP Security

VoIP systems have the same security risks as other IP-based systems. Systems without effective protection can be wiretapped, spit (similar to e-mail spam) can be received, and so forth. If a company started to use VoIP focusing only on financial advantages without knowing the "dark side" of the technology, the financial balance could easily turn to negative.

Currently encryption and authentication of user access is only a recommendation by H.323. (H.323 is the international standard for multimedia communication over packet-switched networks, including LANs, WANs, and the Internet.) This means that any H.323-aware user inside the company can tap into any conversation on the system and any outside person can monitor every conversation with access from her or his desk.

In the case VoIP technology is used for a remote access location, another type of serious security risk arises because of problems with firewalls. H.323's firewall negotiation mechanisms require direct access into the corporate network, so the company has to open its entire network up to all UDP and TCP traffic. The solution is to use an H.323-aware firewall.

Security is one issue that most companies have take into account when implementing new communication systems and this is a good reason for gradually introducing new technology.

Security Technologies for Wireless Communication

Wireless Wide Area Networks (WWAN)

Mobile Security

Mobile security is inherently different from LAN-based security. The basic demands for privacy (confidentiality), integrity, authenticity, and non-repudiation are even harder, as the range of users is broader than in traditional networks. As security in the mobile world is more complex and different, it needs more advanced network security models. It can state that mobile communication is one of the biggest changes in the security market. Mobile security measures depend on the types of data and applications being mobilized. The more sensitive the data, the more strict security measures must be introduced.

Enterprises must be aware of how traditional security challenges change in relevance in a mobile world. Some special considerations for mobile security include the following:

- **Problem of authentication.** As companies report very high numbers of mobile device theft or loss, simply authenticating the mobile device is insufficient. A process of "Two Factor Authentication" had to be introduced. This technology is used to verify both the device and the identity of the end-user during a secure transaction (i.e., two-factor authentication confirms that both the device and the user are authorized agents). Two-factor authentication is critical in protecting network integrity from the inevitability of stolen or lost devices.
- **Minimize end user requirements.** End users are impatient when using mobile services. They want access to applications and data immediately and will resist time-consuming accessing tasks. Requiring end users to conduct complex security processes is counterproductive to the purpose of mobile computing, and further exposes the enterprise to security breach. While a successful mobile application will require some user

participation, involvement should be restricted to quick, easy, and mandatory tasks. Password-protect enterprise applications, an alternative to power-on password authentication, require users to enter a password or pen-based signature when accessing company content. This is a critical first-step in mobile security procedures.

It is critical that a mobile application supports industry-standard security protocols, including:

- **SHTTP:** This is Secure Hypertext Transport Protocol run on a Secure Socket Layer (SSL).
- **WTLS:** Standard for Wireless Transport Layer Security. This protocol provides authentication and encryption for WAP devices.
- **WPKI:** WAP PKI (used by VeriSign) to maintain security. PKI provides an infrastructure and procedures required to enable trusted partnerships needed to authenticate servers and clients in wireless application environments.
- **Any type of standard encryption technology:** e.g., RSA, Triple DES.

Implement WPKI authentication technology. PKI, or Public Key Infrastructure, is a protocol-enabling digital certificates on wired devices. WPKI is an adaptation of PKI for mobile devices that meets m-commerce security requirements. Because PKI functions are bandwidth intensive and require processors tuned expressly for PKI operations, using a PKI proxy server allows balancing processing between the mobile device, the mobile application server, and the proxy server.

WTLS. WAP includes the wireless transport layer security (WTLS) specification, which defines how Internet security is extended to the mobile Internet. WTLS is poised to do for the wireless Internet what SSL did for the Internet—open whole new markets to m-commerce opportunities.

There are three steps to the WAP security model:

- WAP gateway simply uses SSL to communicate securely with a Web server, ensuring privacy, integrity, and server authenticity.

- WAP gateway takes SSL-encrypted messages from the Web and translates them for transmission over wireless networks using WAP's WTLS security protocol.
- Messages from the mobile device to the Web server are likewise converted from WTLS to SSL. In essence, the WAP gateway is a bridge between the WTLS and SSL security protocols.

WTLS was specifically designed to conduct secure transactions without requiring desktop levels of processing power and memory in the mobile device. WTLS processes security algorithms faster by minimizing protocol overhead and enables more data compression than traditional SSL solutions. As a result, WTLS can perform security well within the constraints of a wireless network. These optimizations mean that smaller, portable consumer devices can now communicate securely over the Internet. The translation between SSL and WTLS takes milliseconds and occurs in the memory of the WAP gateway, allowing for a virtual, secure connection between the two protocols.

FLASH-OFDM Security

FLASH-OFDM has ideal characteristics for VPN application. It has broadband speed, low-latency connectivity, authenticated access, and full mobility. As the application of VPN decreases the downlink rate approximately 25-30%, there is enough reserve in the technology. Its end-to-end latency of 35 milliseconds enables the timely LAN-VPN synchronization so critical to the success of the application.

The FLASH-OFDM link layer security can protect the air interface between the wireless device and the network access node. This involves integrity protection, encryption, or both. Also, to address security concerns that affect multiple protocol layers and applications, an enterprise should cost-effectively employ security at the network layer or higher.

Wireless Metropolitan Area Network (WMAN)

Taking into consideration the problems of the WEP in the 802.11 Wi-Fi history, the standardization bodies have prioritized security from the beginning. Therefore, base station designers require a dedicated high performance security processor. The WiMAX standard requires that all traffic must be encrypted with CCMP (Counter Mode with Cipher Block Chaining Message Authentication Code Protocol). CCMP uses AES to provide the encryption for secure transmission as well as data authentication for data integrity.

WiMAX will face the same challenges once certified equipment starts to become available. Although the 802.16 standards have far greater security functionality built into the base than Wi-Fi did, the perception of their safety will have to be high before they win the trust of enterprise and carrier users. As in the case of other standards, many advances will come from individual vendors, whether as enhancements that differentiate an individual product, or work that may be fed back into the standards process.

There are security processors that have been specially developed for WiMAX offering a suitable encryption and security solution for the evolving 802.16e standard. The symmetric key cores, which include the ability to perform AES-CCM function, perform at 200 Mbps with 1500-byte packets. It is this small packet performance, coupled with the internal 32Kb of memory, that makes the processor ideal for WiMAX base stations by performing the complex encryption/decryption with minimal latency. For multiple-channel base stations a processor can perform AES-CCM with 1500 Byte packets at 275 Mbps.

The best processors are designed for advanced networking applications like virtual private networking (VPN) broadband routers, wireless access points, VPN edge router/gateways, firewall/VPN appliances, and other network and customer premise equipment. Some of them can handle a variety of IPsec and SSL/TLS protocols including DES, 3 DES, AES and public key. In addition to IPsec and SSL protocols, the temporal key integrity protocol (TKIP) and AES counter mode encryption can be also supported.

Wireless Local Area Networks (WLAN)

A user of the wireless network can apply a variety of simple security procedures to protect the Wi-Fi connection. These include enabling 64-bit or 128-bit Wi-Fi encryption (Wired Equivalent Privacy-WEP), changing the password or network name and closing the network. These basic techniques work in both small offices and large corporations. However, additional, more sophisticated technologies and techniques can also be employed to further secure the business network (WI-FI Alliance, 2004).

WEP and other wireless encryption methods operate strictly between the Wi-Fi computer and the Wi-Fi access point or gateway. When data reaches the access point or gateway, it is unencrypted and unprotected while it is being transmitted out on the public Internet to its destination—unless it is also encrypted at the source with SSL or when using a VPN (Virtual Private Network). WEP protects the user from most external intruders, but to reach a more secure connection additional technologies have to be applied, as WEP also has known security holes. There are several technologies available, but currently the VPN works best.

There are other security technologies that can apply for WI-FI. Kerberos—Another way to protect the wireless data is by using a technology called Kerberos. Created by MIT, Kerberos is a network authentication system based on key distribution. It allows entities to communicate over a wired or wireless network to prove their identity to each other while preventing eavesdropping or replay attacks. It also provides for data stream integrity (detection of modification) and secrecy (preventing unauthorized reading) using cryptography systems such as DES.

The Media Access Control (MAC) Filtering—As part of the 802.11b standard, every Wi-Fi radio has its unique Media Access Control (MAC) number allocated by the manufacturer. To increase wireless network security, it is possible for an IT manager to program a corporate Wi-Fi access point to accept only certain MAC addresses and filter out all others.

The RADIUS (Remote Access Dial-Up User Service) Authentication and Authorization is another standard technology that is already in use by many companies to protect access to wireless networks. RADIUS is a user name and password scheme that enables only approved users to access the network; it does not affect or encrypt data.

Because of the extraordinary success and adoption of Wi-Fi networks, many other security technologies have been developed and are under development. Security is a constant challenge, and there are thousands of companies developing different solutions. There are a variety of security solutions that effectively are put on the "top" of the standard Wi-Fi transmission and provide encryption, firewall, and authentication services. Many Wi-Fi manufacturers have also developed proprietary encryption technologies that greatly enhance basic Wi-Fi security.

An important problem is the Wi-Fi Security in public spaces. Wireless networks in public areas and "Hot Spots" like Internet cafes may not provide any security. Although some service providers do provide this with their custom software, many Hot Spots leave all security turned off to make it easier to access and get on the network in the first place. If security is important for the user the best way to achieve this when one is connecting back to the office is to use a VPN. In case the user does not have access to a VPN and security is important, it is better to limit the use of wireless network in these areas to noncritical e-mail and basic Internet surfing.

Individuals and companies that have the need to go beyond basic security mechanisms can choose to implement and combine these basic technologies to increase protection for their mobile workers and their data. As with any network, wired or wireless, the more layers of security that are added, the more secure the transmissions can be.

Wireless Personal Area (or Pico) Network (WPAN)

Bluetooth Security

As Bluetooth is a relatively new technology and attacks on Bluetooth devices at this stage are relatively new to consumers, the attacks are not widely seen as a real threat. But for today it becomes clear that there are sensitive information on both company and personal levels that can be accessed relatively easily. So, it is important to know what types of information are in danger, how attacks can be committed, and how to fight against them (Gehrmann, Persson, & Smeets, 2004).

By Bluetooth attacks the following confidential data are in danger:

- In the case of mobile phones the entire phone book, calendar and the phone's IMEI.
- Complete memory contents of some mobile phones can be accessed by a previously trusted ("paired") device that has since been removed from the trusted list.
- Access can be gained to the AT command set of the device, giving full access to the higher level commands and channels, such as data, voice, and messaging.

When Bluetooth are included in laptop PCs this raises the possibility of opening a wireless back door into all data stored on the PCs.

The main attack types and their possible results are as follows:

- The "Snarf" attack—It is possible for attackers to connect to the device without alerting the user, and once in the system sensitive data can be retrieved, such as the phone book, business cards, images, messages, and voice messages.
- The "Backdoor" attack allows attackers to establish a trust relationship through the "pairing" mechanism, but ensuring that the user cannot see the target's register of paired devices. In doing this attackers have access to all the data on the device, as well as access to use the modem or internet; WAP and GPRS gateways may also be accessed without the owner's knowledge or consent.
- The "Bluebug" attack—This attack gives access to the AT command set; in other words, it allows the attacker to make premium-priced phone calls, allows the use of SMS, or connection the Internet. Attackers cannot only use the device for such fraudulent exercises—it also allows identity theft by impersonating the user.
- "Bluejacking" allows attackers to send messages to strangers in public via Bluetooth. When the phones 'pair' the attacker can write a message to the user. Once connected the attacker may then have access to any data on the user's Bluetooth device, which has obvious concerns.
- Phones are vulnerable when they are in "discoverable" or "visible" mode, and the Bluetooth functionality is enabled. Visible mode lets Bluetooth phones find other Bluetooth phones in their vicinity so phone owners can exchange electronic contact information. Users can turn the visible mode off, but some models can be attacked even when a user turns off the visible mode. There are lists of Bluetooth devices on the Internet that give the security level of each device.

RFID Security

When talking about security it should be taken into account that the primary purpose of the RFID technology is the realization of cheap and automated identification. So, standard security mechanisms cannot be implemented because of their relative complexity compared to the limited computing resources of a tag. Cryptography like DES, AES, or efficient public-key protocols is too memory-consuming for low-cost tags.

Recently, breaking tags and disturbing the identification systems became "popular," so the security of RFID systems came into focus. The communication between reader and tag is unprotected in most cases, so eavesdroppers may thus listen in if they are in immediate vicinity. The tag's memory can be read if access control is not implemented. The attacks against RFID systems can be grouped according to which security service is under fire (Knospe, 2004):

- With the exception of high-end ISO 14443 systems, which use message authentication codes (MACs), the integrity of transmitted information cannot be assured. Checksums (CRCs) are often employed on the communication interface but protect only against random failures. Furthermore, the writeable tag memory can be manipulated if access control is not implemented.
- Any RFID system can easily be disturbed by frequency jamming. But, denial-of-service attacks are also feasible on higher communication layers. The so-called "RFID Blocker" disturbs the communication of a reader with all or with specific tags to protect the privacy of consumers.

RFID systems are already used for a large number of applications related to object identification. But, there remain still a number of issues to be resolved. Open technical issues are related to tag orientation, reader coordination, and the relatively short range, to name a few. Furthermore, a number of security and in particular privacy questions are still open. Consumer concerns may form an obstacle to further commercial deployment. Although today sophisticated mechanisms cannot be implemented on a 5-cent tag, a number of proposals exist even for very restricted resources.

Interoperability in Smart Organizations

Importance of Interoperability

There is an extremely big number of different information systems operating in connecting fields all around the world. Many of them are or should be connected to other systems somehow, as data, information, or knowledge exchange would be needed. The speed and reliability of this exchange is also important. Interoperability deals with the solution of this problem. Interoperability is the ability of two or more systems or components to exchange information and to use the information that has been exchanged (IEEE, 1990).

Interoperability is the capability to provide successful communication between end-users across a mixed environment of different domains, networks, facilities, equipment, and so forth from different manufacturers and/or providers. In this context, the communication is meant between end-users or between an end-user and a service provider.

Interoperability is a real and expensive problem. A study of National Institute of Standards and Technology (NIST, 1999) reported that the U.S. automotive sector alone expends one billion dollars per year to resolve interoperability problems. Half of this cost was spent on data file exchange issues.

With a good solution millions of dollars can be saved. According to a study of NIST (NIST, 2002) with the application of the most widely used component of the STEP (Standard for the Exchange of Product Model Data) standard (ISO 10303-203 ([SO 1994b], Configuration controlled design) only the transportation equipment manufacturing community saved over $150 million per year. This figure is expected to rise to $700 million by 2010.

The STEP standard became so important and popular that an international organization, the International Alliance for Interoperability (IAI) has been founded to make the new results even more effective (IAI, 2005).

Levels of Interoperability

The exchange of data and information can happen in a very wide range. There is a need to give some type of classification. As interoperability is a practical problem and users and enterprises had to solve these types of problems on their own, so the classifications have evolved according to professional fields and communities. Interoperability is about information exchange, so these questions arise when connecting two or more systems. When approaching this connecting activity from the other side, experts are talking about system integration. There are different approaches to integration according to the object and field of integration.

Interoperability can be qualified at different levels in networking (e.g., protocol interoperability, service interoperability). It can be achieved through various types of interworking and interconnection:

1. Network interworking: interactions between different types of networks, end-systems, or parts thereof, with the aim of providing an end-to-end communication for a specific service;

2. Service interworking;
3. Terminal and peripheral interworking and interconnection.

In the field of manufacturing and design the integration of the application systems of the product model, the CAD/CAM systems can be done at two levels (Eversheim, 1991):

1. Data/information integration through data/information exchange between CAD and CAM applications without loosing intention, content, and so forth. Tools: IGES, SET, STEP, etc.
2. Functional integration through the communication between CAD and CAM applications. This communication organizes and links the various functional areas. Integrating all aspects of the information involved in a product life cycle into a single shared information model. Tools: product model, manufacturing model, enterprise models.

In case of smart organizations, the three-level integration/interoperability hierarchy can be applied, originally developed for enterprise integration by the AMICE Consortium (AMICE, 1993).

1. Physical, or System Integration

Integration aspects: basic computer communication, message passing, and interconnection. Applied technologies: OMG/CORBA, TCP/IP, Internet/WWW.

2. Application Integration

Integration aspects: distributed co-operative applications, data/information exchange, and application interoperability. Applied technologies: EDIFACT, STEP/PDES, KIF/KQML, HTML.

There are different big projects, development in progress or already finished in this field. In the followings some of them are introduced briefly:

TMF/NGOSS and OMG/MDA

The mission of Object Management Group (OMG) is to help computer users solve integration problems by supplying open, vendor neutral interoperability specifications. The Model Driven Architecture™(MDA™ is OMG's current strategy in solving integration problems.

The NGOSS (New Generation Operations Software and Systems) architecture of TeleManagement Forum (TMF) is described using technology-neutral constructs. These include concepts taken from RM-ODP as well as extensions to the basic UML metamodel to represent fundamental NGOSS concepts and principles. It does not prescribe a single new technology—rather, it allows for a federation of different technological approaches, each of which offers particular advantages at the business and system levels

Some of the main goals of TMF/NGOSS and OMG/MDA work are to provide benefits through technology neutral architecture; that is, architecture that is sustainable through technology changes, cost effective application integration, such as interworking and interoperability through application of standards (Faurer, Fleck, Huang, Richardson, & Strassner, 2004).

NIIIP (National Industrial Information Infrastructure Protocols)

In the case of virtual enterprises, the National Industrial Information Infrastructure Protocols (NIIIP) Reference Architecture (NIST, 1998) has been developed for global information exchange and sharing by the NIST. This architecture is one of the most complete realizations of virtual enterprise architectures. It intends to bring together the product realization process integration efforts, by developing general global protocols for the technical standards of product data definition, communication, and object technology and workflow management. The NIIIP doesn't intend to develop a new system, but rather apply existing standards to consolidate, harmonies, and integrate the many sets of existing protocols. The main goals of the NIIIP reference architecture is to help the establishing and operating of VEs in the industry by applying standardized solutions for VE connectivity, for industrial information modeling and exchange and management of VE projects and tasks. NIIIP defines a series of protocols that make the STEP-defined data available as data objects in an OMG CORBA environment.

KIF (Knowledge Interchange Format)

The application of knowledge-based systems become more frequent, so the knowledge exchange and knowledge sharing have an increasing role. In this field, KIF (Knowledge Interchange Format) is a language designed for use in the interchange of knowledge among disparate computer systems (Genesereth,

1992). It has declarative semantics (i.e., the meaning of expressions in the representation can be understood without appeal to an interpreter for manipulating those expressions); it is logically comprehensive (i.e., it provides for the expression of arbitrary sentences in the first-order predicate calculus); and it provides for the representation of knowledge about knowledge.

PIF (Process Interchange Format)

The PIF (Process Interchange Format) serves to exchange process information (Lee, 1994). The goal of PIF is to offer an interchange format that helps automatically exchange process descriptions among a wide variety of business process modeling and support systems, such as workflow software, flow charting tools, process simulation systems, and process repositories.

3. Business Integration

Integration aspects: business process coordination, enterprise-wide knowledge sharing, interworking. Applied technologies: CIMOSA, GERAM, ENV 40003.

AMIS (Automated Methods for Integrating Systems) Project

In the field of manufacturing automation the NIST has started the AMIS project (Libes et al., 2004).

There are three main areas of work in the AMIS project: interaction ontology formulation, semantic mapping, and connector transformation. Interaction ontology formation is concerned with capturing the "business" and "engineering" interaction concerns in a form suitable for reasoning. Semantic mapping pertains to building tools to create semantic maps among ontologies. Connector transformation is concerned with creating generators for dynamic message converters; this will ultimately expand to support dynamic protocol conversion. Efforts in each of these areas must come together to support automation in the integration process.

Systems engineers use a combination of top-down and bottom-up approaches to match business process objectives with component functionality. The AMIS approach is to formalize and capture the information the system engineer uses to perform this matching and then use software-based reasoning tools to support automation of the integration task.

Impacts of Wireless Technologies on Smart Organizations

Fields of Impacts

According to different studies, the introduction of new ICT to firms can indicate changes generally in organizational structure, in the competitive strategies, and environment (appearance of new products and services). Other studies declare that the introduction of new ICT and the organizational changes are designed to achieve greater productivity and flexibility. The conclusion of the study (Bocquet & Brossard, 2003) is (probably it is the most realistic) that there is no general model how ICT effects a company; the extent, the fields, and type of changes vary from company to company. So, all the impacts introduced in this chapter usually cannot be detected in one enterprise, and impacts not listed here can be found in other cases.

The new networking technologies extend company data, back-end information systems, and e-mail to mobile employees, broadens the accessibility of mission critical data. Mobile access modifies the way workers interact with colleagues, customers, and suppliers.

The possible fields of impact introducing wireless communication systems are:

- organizational structures
- SW and HW elements
- physical elements in administration/production
- human resources
- working environment—office structure, equipments for people, physical env.
- workflow
- business processes
- speed of communication, the validity/availability of information, reliability of communication, security (trust), cost of implementation, cost of reconfiguration (on shop floor level)
- market strategy
- management.

Main Impacts

The first big change for enterprise information systems was caused by the application of Internet, the introduction of the global network. The next step forward was caused by the WEB-based technologies and to a lesser extent the mobile technology also has appeared on the scene. The latest effect on organizations (both in structures and operation) is caused by the different advanced wireless technologies. The result of their integration can be really called "cyberspace."

Mobile technology affects both the structure and the operation of enterprises. The main reasons to develop a mobile solution in the organization can be to provide access to company e-mail and to intranet applications, to develop specific company applications, to keep permanent contact with service workers, to improve work scheduling, and offer possibility for m-commerce.

The interactions among organizations become easier and more transparent. Wireless technologies simplify the processes of maintaining visibility and control over transactions within a networked organization and allow real-time collaboration. The change from wired to wireless technology is probably even harder than the change in the previous cases, as the application of these technologies and means (e.g., mobile phones) alters not only work processes but the social and cultural environments as well.

The listed influences are general ones, as their effects can vary according to the application field (service company, automated manufacturing company using sensors, etc.), size of the organization, the cultural environment, and many other factors. In the following, only the main areas of impacts will be presented.

Impacts on Organization (Structure and Work Processes)

- Organizational structure has to change to flat, open, lean structure with a 24-hour 365-day availability.
- Traditional functional and hierarchical barriers have to be eliminated while supporting teamwork and open access to people and information.
- Clear responsibilities for basic and auxiliary activities.
- The whole organization has to change to customer-oriented—real-time information collaboration among the participants. Benefits: innovative and agile work process.

Impacts on Working Technology

- It is important to separate basic/core activities from auxiliary activities. A basic activity is one that contributes to the competitive advantage of the organization. Auxiliary activities don't contribute directly to the competitive edge (e.g., administration). The right selection and the balance of the two activities are essential for the organization (Sifonis, 2003).
- The significance of team-work is growing.
- Remote meetings of different groups on different levels help collaboration.

Impacts on Information Technology

- Besides standard word processing and spreadsheet applications, groupware technologies are also included which require not only new technical skills but the development of new ways of working.
- Technologies such as shared databases and a calendaring and scheduling tool support asynchronous group work, networked electronic whiteboard (which allows for shared viewing and manipulation of files between multiple sites), remote video linkages across multiple sites, and desktop conferencing (joint use of a single application running on two desktops) support synchronous group work.
- The different databases have an outstanding role. The various databases encompass routine and non-routine work, and form a kind of glue that holds different groups together.
- Secure communication is a very important demand in a system continuously connected, as in case of wireless networks, so WNs need stronger security as wired systems (e.g., VPN). This enhanced security demand needs modification in system infrastructure and in architecture as well. Other security related services like access right structure and archiving of documents also has to be modified.

Interoperability

The development of wireless technologies is very fast, and there are solutions that did not exist a year ago. These technologies converge data, voice, graphics, and video over a single network, and they allow each member of the network to access them without any space and time limitation. The representation and exchange form all of these descriptions have to be standardized, or else their access would not be possible for the different systems.

Impacts on Human Resources

- For the management, it is important to redefine authorities and responsibilities more clearly.
- Organizing remote meetings on different levels helps to better solve integrated problems, and these videoconferences also help trust building among the teams and members of the organization.
- For the staff, continuous training is important in order to use the new technology effectively, so for them the motivation for learning is a basic must.

Future Trends in Communication of Smart Organizations

The development and breakthrough of wireless communication is extremely fast, so today it can be seen what happens when a society, economy, and networked communication goes unwired. The application of mobile/wireless equipment of different kinds is dramatically increasing, and new technologies are appearing. The effects of this mobility are also remarkable in working communities and enterprises.

The basic trend in wireless communication technology is the convergence and integration of the different technologies and the efforts to increase the speed of data transmission rates. An additional reason for integration is to extend the coverage; that is, to increase the geographical availability and reach. Finally, the decrease of service costs is a general requirement of the market. Of course, these are visible technical and economical tasks for the wireless technology; in the background huge investments have to be done to realize these objectives that are important for the users.

Some examples of the efforts to increase speed and covered area include the present effort to develop the fourth generation (4G) mobile communication, which will have higher data transmission rates than 3G and are planned to be up to 20 megabits per second (about 10 times faster than top transmission rates planned for 3G mobile) (Gupta, 2004). In principle, 4G will allow high-quality, smooth video transmission. Another result of the developments is the dual-mode GSM/satellite phones that are switching automatically according to the covered area using the more economic operation mode.

Parallel with the integration of existing technologies, big efforts are invested into the development and standardization of new wireless technologies (e.g., WiMAX, FLESH-OFDM, UWB). The IP-based fixed and wireless broadband technologies can significantly redraw the world of communication.

WLAN and 2.5/3G can be considered as complementary technologies. Wi-Fi offers great bandwidth in close proximity to a base station, but has limitations as users move out of range, and currently there aren't enough Wi-Fi base stations deployed publicly to support ubiquitous access. GPRS technologies like 2.5 can support speeds that beat dial-up from practically anywhere in the network's coverage area—which in most cases is a substantial range. Manufacturers integrate their data connect product with 802.11 technology—in this case, it is a GPRS-based PC Card for laptops and PDAs.

The wireless network of the future will be a hybrid solution of WWAN/WLAN/WPAN (e.g., 3G/Wi-Fi/Bluetooth/UWB), technologies with roaming and billing systems that provide the bridge. This "wireless chain" will offer continuous Wi-Fi coverage in dense metro areas and 2.5G or 3G in more outlying areas (Vaughan-Nichols, 2003). Wireless systems will become pervasive and will exist in a multitude of flavors (sensors, satellites, LANs, PANs, cellular, access, etc.).

Different wireless communication technologies are developing very fast. From a practical aspect this means that their integration will be holistic; that is, the user can move anywhere smoothly, with full broadband availability and without realizing that he or she changes the different types of network. The user's body area network (BAN) moves together like a "bubble" through the space.

This level and type of availability will result in additional changes in society and all sectors of economy. Smart organizations can operate more effectively with this information infrastructure, unless there will be new production and service organization structures and smart organization will be history only.

In spite of the many positive characteristics and effects introduced so far, mobile communication has negative sides as well; for example, the possibility of tracking services and routes of owners by agencies and competitors and marketers using information from wireless devices. The same can be applied for RFID technology, so it is not suggested to use this technology in certain states because of privacy reasons. A general problem is that wireless communication is more sensitive for illegal wire-tapping and in some cases (e.g., Wi-Fi) the security of networks is not guaranteed when using improper security packages.

Of course, it is impossible to predict exactly the technology developments and the evolution of culture and customer needs, but customers do not really care about which technology is used, but about cost and speed. Indeed, speed, cost, and coverage are what are likely to decide whether one technology will triumph or the other or whether they'll work best for users together.

Conclusion

Network-based organizations, like smart organizations, are the main components of the information and knowledge society. The market competition force networked organizations in the direction of continuous change to be able to follow the frequently changing market demands.

These organizations apply ICT very intensively, both for internal and external communication, supporting their cooperation in order to react flexibly to changing business environment. As wireless information and communication technologies make their operation more effective both in the fields of production and finance, the competitiveness of networked production systems is increased through wireless networks.

The infocom systems applied by the companies have their human part (users) as well, so the importance of trust is increasing very quickly. As it is pointed out by different analyses based on real-life statistics, when users do not trust an IC system or service they do not use it. Security services partially provide this trust for the users, while the human side of trust depends on the media of communication, the structure of interfaces, and on the duration of contacts. The organizations have to adapt their IC systems to these requirements as well, even by slightly changing their culture or organizational structures.

The different types of wireless communication technologies originating from their openness and flexibility will be always a security risk. The managers of information technology have to adapt those technologies, tools, and devices into their systems that can induce high trust-level in all humans involved in different levels of the smart organizations.

The new generations of networking technologies, the different types of wireless communication systems make significant changes not only in the operation of networked-based organizations, but in the cultural and social environments as well. The paper introduced briefly the main groups of wireless networks and their effects on networked organizations. The main conclusion is that the wireless networking technologies can cause remarkable modifications in the structure, in the operation, in the collaboration techniques, in the cost structure, and in business processes of any type of organizations, so their application has strategic importance for all types of companies.

References

AMICE. (1993). *CIMOSA: Open System Architecture for CIM* (2nd revised and extended version). Berlin: Springer Verlag.

Anderson, C. (2003). The Wi-Fi Revolution [Special Issue]. *Wired Magazine, 11*(5).

Anderson, R. (2001). *Security engineering: A guide to building dependable distributed systems.* New York: John Wiley & Sons.

Bakker, D. M., & McMichael Gilster, D. (2002). *Bluetooth end to end.* John Wiley & Sons.

Bocquet, R. & Brossard, O. (2003, November 7-10). Determinants of IT adoption, organizational change and the nature of the firm: An empirical study of French survey data, EAEPE 2003. *The Information Society, Understanding Its Institutions Interdisciplinary* (pp. 30-31). Maastricht, The Netherlands, in Book of Abstracts.

Bos, N. D., Olson, J. S., Gergle, D., Olson, G. M., & Wright, Z. (2002). Effects of four computer-mediated channels on trust development. *Proceedings of CHI 2002.* New York: ACM Press.

Castelfranchi, C., & Yao-Hua, T. (Eds.). (2001). *Trust and deception in virtual societies.* Kluwer Academic Publishers.

Cheskin, (1999). *eCommerce trust: A joint research study with Studio Archetype*. Sapient and Cheskin, Retrieved December 15, 2004, from http://www.cheskin.com/p/ar.asp?mlid=7&arid=10&art=0

D'Hertefelt, S. (2000). *Trust and the perception of security*. Retrieved December 15, 2004, from http://www.interactionarchitect.com/research/report20000103shd.htm

Elbert, B. R. (2004). *The satellite communication applications handbook* (2nd ed.). Boston; London: Artech House, Inc.

Engst, A., & Fleishman, G. (2003). *The wireless networking starter kit*. Berkeley, CA: Peachpit Press.

Eversheim, W., &Gross, M. (1991, May). Trends and experiences in applying simultaneous engineering. *Proceedings of the Seventh CIM-Europe Annual Conference* (pp. 3-11).

Faurer, C., Fleck, J., Huang, J., Richardson, T., & Strassner, J. (2004). *TMF White Paper on NGOSS and MDA*. Retrieved December 15, 2004, from http://www.tmforum.org/browse.asp?catID=1705&linkID=28796

FBI (2004). *The 2004 CSI/FBI Computer Crime and Security*. Vol. IX. No. 1. Retrieved December 15, 2004, from http://www.gocsi.com/forms/fbi/pdf.html

Filos, E., & Banahan, E. (2000). Will the organisation disappear? The challenges of the new economy and future perspectives. In L. M. Camarinha-Matos, H Afsarmanesh, & R. J. Rabelo (Eds.), *E-business & virtual enterprises* (pp. 3-20). Dordrecht: Kluwer.

Finkenzeller, K. (2003). *RFID Handbook: Fundamentals and Applications in Contactless Smart Cards and Identification* (2nd ed.). John Wiley & Sons.

Flarion Technologies, Inc. (2004). *The benefits of a packet-switched, all-IP mobile broadband network*. White paper, February.

Foster, I. (2000). Internet computing and the emerging grid. *Nature, 7*, December.

Foster, Kesselman, & Tuecke. (2000). *The anatomy of the grid—Enabling scalable virtual organisations*. (2000, White Paper).

Fukuyama, F. (1995). *Trust—The social virtues and the creation of prosperity*. New York: The Free Press.

Gartner Symposium/Itxpo. (2003, October). Lake Buena Vista, Florida, USA, 19-24.

Gehrmann, C., Persson, J., & Smeets, B. (2004). *Bluetooth security*. Boston: Artech House, Inc.

Genesereth, M. R., & Fikes, R. E. (1992). *Knowledge interchange format, Version 3.0 reference manual* (Tech. Rep. Logic-92-1). Stanford University, Computer Science Department.

Gupta, P. (2004). Mobile wireless communications tomorrow. In *Wireless Developer Network*. Retrieved December 15, 2004, from http://www.wirelessdevnet.com/channels/wireless/training/mobilewirelesstomorrow.html

Handy, C. (1995). Trust and the virtual organization. *Harvard Business Review, 73*(3), 40-50.

Hübscher, B. (2004). *Highspeed Internet on the power grid*. Retrieved November 10, 2004, from http://www.ascom-astel.com.ua/plc_ovr.pdf,

Institute of Electrical and Electronics Engineers (IEEE). (1990). *IEEE Standard Computer Dictionary: A Compilation of IEEE Standard Computer Glossaries*. New York.

International Alliance for Interoperability (IAI). (2004). http://www.iai-international.org/

ISO 10303-203. (1994). Industrial automation systems and intregration — Product data representation and exchange — Part 203: Application Protocol: Configuration controlled design. International Organization for Standardization, Geneva, Switzerland.

ISO/IEC 10181-1. (1996). Information technology—Open Systems Interconnection — Security frameworks for open systems: Overview.

ISO/IEC 15408. (1999). Evaluation Criteria for Information Technology Security.

Jarvenpaa, S. L., & D. E. Leidner. (1999). Communication and trust in global virtual teams. *Organization Science, 10*(6), 791-815.

Karayannis, G. (2003, December). Standards-based wireless networking alternatives. *Sensors Online*. Retrieved December 15, 2004, from http://www.sensorsmag.com/articles/1203/26/main.shtml

Kelland, B. (2003, January). *Ultra-wideband wireless technology*. Paper presented at 3rd Annual CM316 Conference on Multimedia Systems, Southampton University, UK.

Knospe, H., & Pohl, H. (2004). RFID Security, Information Security Technical Report. Vol. 9, No. 4, pp30-41.

Knospe, H., Pohl, H., (2004). RFID security. *Information Security Technical Report, 9*(4), 30-41.

Koller, L. (2001). Biometrics Get Real. *CardTechnology,* August, 24-32.

Lee, J., Gregg, G., & the PIF Working Group. (1994). The PIF Process Interchange Format and Framework. December 1994. MIT Center For Coordination Science, 1994 WORKING PAPER No. 180.

Libes, D., Barkmeyer, E. J., Denno, P., Flater, D., Steves, M. P., et al. (2004). *The AMIS approach to systems integration: An overview* Tech. Rep. No. NISTIR 7101). National Institute of Standards and Technology, Manufacturing Systems Integration Division.

Lucent Technologies (2003). *3G wireless networks, wi-fi wireless LANs, and secure VPNs: Perspectives for the business on-the-move*. White Paper, March 2003.

Luhman, N. (1979). *Trust and power*. Chichester, UK: Wiley.

Menezes, A. P., van Oorschot, & Vanstone, S. (1996). *Handbook of Applied Cryptography*. CRC Press.

NIST. (1998). NIIIP Reference Architecture Books 0-6. RA 3.1.

NIST. (1999). *Interoperability cost analysis of the U.S. automotive supply chain* (Planning Report #99-1). Retrieved December 15, 2004, from http://www.nist.gov/director/prog-ofc/report99-1.pdf

NIST. (2002). *Economic impact assessment of the international standard for the exchange of product model data (STEP) in transportation equipment industries* (Planning Report #02-5). Retrieved December 15, 2004, from http://www.nist.gov/director/prog-ofc/report02-5.pdf

Orange book. (1985). Trusted computer system evaluation criteria DoD 5200.28-STD. Department of Defense, December 26, Revision: 1.1 Date: 95/07/14.

Rocco, E., Finholt, T. A., Hofer, E. C., & Herbsleb, J. D. (2001, April). *Out of sight, short of trust.* Presentation at the Founding Conference of the European Academy of Management. Barcelona, Spain.

Rousseau, D. M., Sitkin, S. B., Burt, R. S., & Camerer, C. (1998). Not so different after all: A cross-discipline view of trust. *Academy of Management Review, 23*(3), 393-404.

Schneier, B. (1996). *Applied Cryptography*. John Wiley & Sons.

Sheriff, R. E. & Fun, H. Y. (2001). *Mobile satellite communication networks*. UK: John Wiley & Sons.

Sifonis, J. (2003, March/April). The networked virtual organization—A business model for today's uncertain environment. *iQ Magazine*.

Tanenbaum, A. S. (1996). *Computer networks* (3rd ed). Prentice-Hall.

Ungson, G. R., & Trudel J. D. (1999, May). *The emerging knowledge-based economy.* IEEE Spectrum.

Vaughan-Nichols, S. J. (2003, June 20). Wi-fi and 3G together? *Wi-Fi Planet.* Retrieved December 15, 2004, from http://www.wi-fiplanet.com/columns/article.php/2225411

Wi-Fi Alliance. (2004). *WI-FI security*. http://www.weca.net/

Zheng, J., Veinott, E, Bos, N., Olson, J. S., Gary, Olson, G. M. (2002). Trust without touch: Jumpstarting long-distance trust with initial social activities. *Proceedings of the SIGCHI conference on Human factors in computing systems,* Minneapolis, Minnesota (pp. 141-146).

Chapter VIII

Infrastructure Support for Smart Organizations: Integration of Web Service Partners in Heterogeneous Environments

Peter Bertok, Royal Melbourne Institute of Technology, Australia

Xinjian Xu, Royal Melbourne Institute of Technology, Australia

Abstract

In a rapidly changing world, continuous adoption of new practices is crucial for survival; organizations embracing the latest technologies have a competitive edge. Smart organizations readily take on board new organizational forms and practices, those in particular that offer agility and responsiveness. The Internet and the World Wide Web offer a new way of collaboration via Web services, but heterogeneity of different service components make cooperation difficult. This chapter describes a new approach to combine Web services by employing a layered structure, in which composition of a value-added service can be built from individual components, and each service component can have semantically equivalent but syntactically different alternatives.

Introduction

Smart organizations make good use of internal and external expertise. Knowledge existing within the company can be utilized across different departments, and skills available externally can be employed to improve product or service offerings and market position. To gain the most out of collaboration and cooperation across different organizational units, they have to be managed well and should be built on sound infrastructure.

In the last decade, Web services emerged as a very efficient tool to manage business processes on a ubiquitous platform. In many cases, there can be several, semantically equivalent components, as many service providers offer similar services and similar content, albeit on different platforms and with different interfaces. While there may not be one method that addresses all challenges of heterogeneity, the approach of introducing an integration layer can often offer a solution.

This chapter presents the field of Web services that enable smart organizations to approach their operation in a new way. We start with explaining the benefits of internetworking and using the World Wide Web for business process cooperation. This is followed by a brief overview of Web services, presented from the aspect of smart organizations. Next, we address the problem of heterogeneity that poses one of the greatest challenges to the technology. Then we present an approach of introducing an integration layer that helps in overcoming the difficulties of heterogeneity and in adopting one particular service from among many similar ones. A practical example illustrates how the method works. The chapter concludes with an evaluation of the proposed model.

Smart Organizations and Web Services

Smart Organizations

The changes in business practices present new challenges to companies; staying in the front line requires continuous innovation and investment. One of the most important competencies for organizations seeking to thrive is the application of smart technology in a smart way. Organizations are called smart if they have a knowledge-driven, internetworked focus helping and promoting dynamic adaptation to new organizational forms and practices. These organizations can create and exploit new opportunities, particularly those presented by the new economy (Filos & Banahan, 2000).

To succeed, companies need more than just domain expertise. The collective intelligence of an organization can be much more, or much less, than that provided by individual executives and other employees (Perkins, 2002). A key issue for smart organizations is collaboration and cooperation. Smart organizations recognize expertise, seek input from specialists and contract out many subtasks, possibly very important ones, to others who are better prepared for a particular job. Smooth teamwork relies on a well-supported underlying communication infrastructure providing links between different players, and the parties involved must be convinced that it is only through collaboration that the goals can be achieved. Companies are discovering that a sound infrastructure is even more important than specific technologies, as the infrastructure enables organizations to develop and implement new applications rapidly and seamlessly. Good and rcliablc infrastructurc also enables new technologies to be easily employed and to be replaced when needed.

Web Services

The Internet and the World Wide Web have become a widely used communication infrastructure, but its potential in business-to-business (B2B) interactions has not been fully realized yet. Workflows have been widely used for coordinated execution of multiple tasks in heterogeneous environments before, and we are now witnessing the automation of process flows by employing Web-based services. The Web has been enhanced; from content provider it has become a service provider. Business functions have become accessible via networks; Web services can encapsulate sets of coherent operations.

Web services are compositions of service providers and service consumers, also called publishers and subscribers (Clark, Fletcher, Hanson, Irani, Waterhouse, & Thelin, 2002). The service provided can be functionality, such as organizing a travel itinerary, or data access, such as viewing or modifying entries in databases. Web services are seen as a major device to reduce costs and eliminate redundancies. For example, in an organization a centrally located Web service can cater for a number of departments that otherwise would be using different packages to do the same job. An organization can also offer

Web services to external clients, to utilize available capacity or provide better access to its services.

Employing Web services can have strategic advantages. On the consumer side, it can provide organizations with additional functionality that puts them in a better place on the market. On the service provider side, companies can offer their services to a wider range of potential customers who could not be reached via traditional means.

Integration, outsourcing, and restructuring can all be translated into financial benefits and improve a company's standing. Two important questions need to be answered, however. The first one is organizational: How can these benefits be realized? The second question is technical: How can Web services operate in a heterogeneous environment, such as the World Wide Web, where participants may be using their own platforms, own interpretation of services, data, and so forth? This chapter looks at the technical side of the problem—that is, how to make Web services work in a diverse setting, and how to combine them to provide new, value-added services.

Web Services in Smart Organizations

Agility is essential for smart organizations, and to support it in today's internetworked environment a new generation of applications and services are needed. Smart organizations utilize available technology and adopt new developments to be able to respond quickly and assertively in the market place. The primary aim of employing technologies is to reinforce customer focus and understand and satisfy customer expectations; improving efficiency is also important. Utilizing efficient human interfaces in a ubiquitous environment, Web services can reach customers and respond almost instantaneously.

Strategic alliances and partnerships provide competitive advantage, and they require seamless integration of resources as well as knowledge sharing between businesses. Workflows have been the traditional approach to the management of business processes; an internetworked environment facilitates collaboration and enables their integration. Web-based workflows provide agility as well as easy adaptability to changing environments.

Responsiveness is a key to success, and it cannot be achieved without proper expertise. A smart organization is knowledge-driven, and uses internetworking to acquire know-how that may be slow or hard to obtain via conventional channels. Sharing existing or newly obtained knowledge is essential for cooperation, and aligning systems requires a uniform approach. Web technology offers a standard platform and fast, readily available infrastructure. Web-based services enable the use of a large knowledge base that extends beyond individual companies' resources.

Web Service Types

Recent times have seen the rapid proliferation of Web services that were developed on various platforms. A closer look reveals that there are two basic types of Web services: They can be resource-oriented or activity-oriented, depending on the application they support (Snell, 2004). Resource-oriented services maintain a set of resources, usually data, and provide operations on the resources. The number of operations is usually limited; they typically include retrieval, modification, creation, and deletion. Services of this type are termed Representational State Transfer (REST) style services. Activity-oriented services focus on the actions, and the resources they act on can be hidden and are not important for the user. There can be a variety of actions; the user only tells the actions to be performed, and the service handles the resources on the user's behalf. This type of services are called SOAP style Web services, the name referring to the Simple Object Access Protocol (SOAP).

The Problem of Heterogeneity

Web services present the advantage of easy workflow management on a ubiquitous platform, which has the promise of collaborations on an unprecedented scale. A major obstacle on that path is the heterogeneity of Web services, most importantly the heterogeneity of interfaces. Considering the extremely diverse application areas of Web services, a uniform approach to connecting Web services is not apparent yet. In addition, there are numerous Web service platforms, as almost every major vendor developed its own: IBM produced WebSphere, Microsoft has .NET, there is Lotus Domino, just to name a few. There are also countless tools that address different aspects of Web service development, such as Stylus Studio for Xquery-based aggregation (stylusstudio, 2004), Netegrity for identity and access management (netegrity, 2004), Midreef SOAPScope for diagnostics for debugging, testing, and tuning (mindreef, 2004), Parasoft SOAPtest for testing (parasoft, 2004). Some organizations already heavily utilize Web services and they also provide tools to access their technology platform and product data (amazon, 2004). Interoperability between different products can be difficult to achieve, and heterogeneity poses a great challenge.

Integration of Web Services

Heterogeneous Environments

Communication between heterogeneous entities is a complex task, over the Internet in particular. The differences range from programming language to business semantics. To counter the differences, recent advancements saw the emergence of extensible markup language (XML)-based languages for communication between business partners. The aim of these standards was to bridge the gap between communicating business applications. Early work done in this field saw the formulation of the Web services description standard. This standardization lead to a variety of specifications and standards. They addressed different aspects of heterogeneity, and worked toward a standardization framework. Providing value-added services by composing new services out of existing ones was the next step, and service composition and aggregation became the topic of interest.

The Web services model uses the Web to automatically invoke processes, and includes specifications and methods for publishing, calling, and executing them. It is built on Web-based protocols, but does not require traditional Web-based tools, such as browsers. A Web service is a business function made available by a service provider on the Internet, and its clients are software applications, or sometimes human users. A Web service, defined as a modular program, is "a unit of business, application, or system functionality that can be accessed over the Internet by using XML messaging" (Tosic, Pagurek, Esfandiari, & Patel, 2002). The essence of a Web service is to build middleware using Internet protocols, so that the service of an application can be discovered and accessed through the Hypertext Transfer Protocol (HTTP). Via Web services, one service can interact and exchange information with other services across enterprise boundaries. With Web service technology, existing business services can be encapsulated and published over the Web, and business-to-business (B2B) interaction through the World Wide Web can be supported. The four steps of engaging and using a Web service are shown in Figure 1. First,

Figure 1. The general process of engaging a Web service, as presented in Booth et al. (2004)

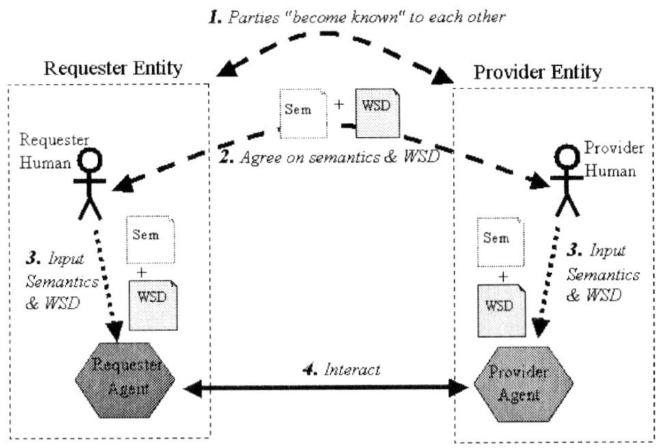

the parties get acquainted with each other. The next step is to find a "common language;" that is, they agree on the semantics and service description of the interaction, and then in step three each of them realizes its own view of the interaction, that is, they implement the requester's service description and semantics and the provider's service description and semantics, respectively. Finally, they exchange messages and the server performs the task or tasks for the requester.

Web services represent the convergence of the concepts of component-based distributed systems (CBDS) and the context of the World Wide Web. To some extent, we can say that the World Wide Web is also a CBDS. Each Web service is a software component in CBDS implementing a specific type of business logic and is independently developed. Like a traditional software component, a Web service component exposes a well-defined, well-documented interface defining the services it provides. Such an interface enables the support of remote access without knowing how the service is implemented (Pires, 2002). In addition, a Web service component has to be robust and reusable (Cicalese & Rotenstreich, 1999). However, we argue that a software component is not necessarily a Web service component. There are several significant differences between them.

- Web service components are self-contained and Internet-enabled (Yang, Papazoglou, & van den Heuvel, 2002). They are accessible through standard Internet protocols such as HTTP, which is vendor independent. In contrast, a software component is accessible through a distributed computing protocol, such as the distributed component object model (DCOM), Common Object Request Broker Architecture (CORBA), or Java Remote Method Invocation (RMI), some of which are vendor dependent (Pires, 2002).

- Web service components are less tightly coupled, as they communicate with each other via XML-based messages, while software components communicate via procedure invocations.
- The connection between Web service components is unplanned. The connection between components is established temporarily in an ad hoc manner because Web service components often operate in a highly dynamic environment (Benatallah, Sheng, & Dumas, 2003).

Due to the adoption of HTTP and XML, Web service components possess a more flexible interoperability than traditional software components.

Web service components are typically built around XML-based standards, such as the Simple Object Access Protocol (SOAP) (Gudgin, Hadley, Mendelsohn, Moreau, & Nielsen, 2003), Web service description language (WSDL) (Christensen, Curbera, Meredith, & Weerawarana, 2001), and universal description discovery and integration (UDDI, 2000). Even though Web service components can be implemented in any programming language on any platform, the service description language is common: WSDL. The services interact with each other via a standard XML messaging protocol, such as SOAP, which guarantees broad interoperability among components. In other words, XML-based standards act as wrappers or bridges to support interoperability between components. In this chapter, we refer to interacting Web service components as service partners. Accordingly, a Web service component can be regarded as a client (requiring services) or a server (providing services). The World Wide Web Consortium defines Web services the following way: "A Web service is a software system designed to support interoperable machine-to-machine interaction over a network. It has an interface described in a machine-processable format, namely in WSDL. Other systems interact with the Web service in a manner prescribed by its description using SOAP messages, typically conveyed using HTTP with an XML serialization in conjunction with other Web-related standards" (Booth et al., 2004).

Web service components can be heterogeneous in many aspects: implemented in different programming languages on different platforms, employing various communication methods, using a variety of interfaces, data formats, and data encoding schemes, having diverse content capabilities, and so forth. Our primary focus here is the heterogeneity of interfaces of Web service components, which can be manifested in capability differences between interfaces and in syntactic differences between interfaces of semantically equivalent components.

Multiple Web service components can be assembled to deliver more valuable services. A large number of enterprises have already put their services on the Web, with an even larger number of Web service components expected to appear in the near future. To provide value-added services, some enterprises may combine their services, and there is a need for mechanisms to integrate Web service components. Integration refers to the aggregation of basic Web service components, possibly provided by different enterprises, and relies on dynamic discovery and execution of those basic Web services. In other words, a value-added service is implemented by the integration of operations of different basic Web services (Benatallah, Dumas, Sheng, & Ngu, 2002). Integration of Web service components is highly dynamic; changing market conditions and business objectives affect business logic and lead to continuous evolution of components and aggregations.

This chapter describes a solution for integration of heterogeneous Web services, which can cater for the dynamic needs of today's business environment. A multiplayer approach is employed, and the modular structure enables easy adaptation or replacement of components.

XML-Based Standards

Interoperability of Web services is ensured by a number of methods, primarily by several standards. The most important relevant standards are described as follows.

- Web service description language (WSDL) introduces a standard way to describe a Web service component. It provides the service requester with information on the functionality a Web service component offers, and on where and how it can be accessed.

The WSDL standard does not define collaboration or interaction between different services.

- Simple object access protocol (SOAP) defines communication and allows a Web service component to send an XML message to another service component. Interoperability is ensured primarily by this protocol.
- Universal description, discovery, and integration (UDDI) offers a platform-independent way to register and discover a specific Web service component. The UDDI server is the first point that a service requester has to contact before accessing a Web service component. Actually, the UDDI itself is also a Web service component that provides registry and discovery services for other Web service components. The UDDI server acts as "yellow pages" to list the available services.

Service-Oriented Architecture

Figure 2 shows how these standards relate to each other in the Web service infrastructure. When a company wants to make one of its services available on the Web, it creates a WSDL description of the service and registers this description with the UDDI registry server, and a pointer to the WSDL

Figure 2. Web service infrastructure

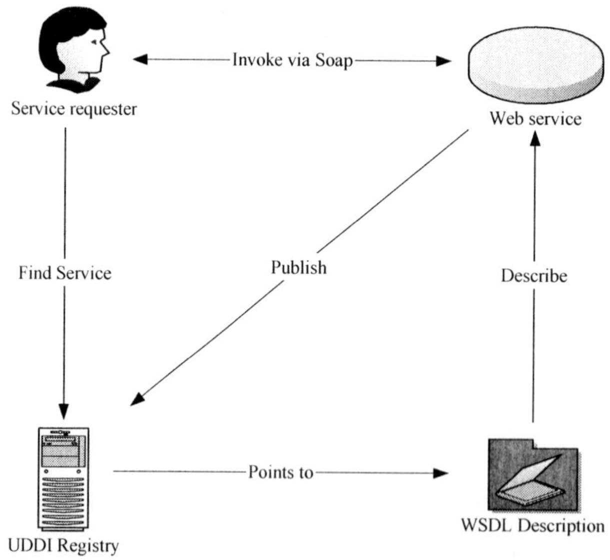

Table 1. Web service protocol Stack (Cerami, 2002; Tilley, Gerdes, Hamilton, Huang, Muller, & Wong, 2002)

Service	Implementation
Discovery	UDDI (Universal Description, Discovery and Integration)
Description	WSDL (Web Service Description Language)
XML Messaging	SOAP (Simple Object Access Protocol)
Transport	HTTP (Hypertext Transfer Protocol)
Network	TCP (Transmission Control Protocol)
	IP (Internet Protocol)

description document is stored at the UDDI registry server. When an application wants to use a particular service, it sends a query to the UDDI registry server to find the service, which matches its needs and retrieves the WSDL description that contains the interface and location of the service. Finally, the service requester creates a SOAP message with which to invoke the service.

Table 1 shows how the different standards build on each other, and how they relate to the TCP/IP network reference model.

Composing and Recomposing Web Services

To improve flexibility, several approaches have been used to enable a system to dynamically adapt to different component compositions. These approaches are classified into two groups, depending on whether existing relationships between composed components are broken or not.

Adjusting relationships between the components without breaking them can be performed in several ways, such as the following (Tosic, Pagurek, Esfandiari, & Patel, 2002):

1. Manipulating the parameters of components to provide variant services.
2. Providing multiple classes of services.
3. Recustomizing the services according to user profiles.
4. A dedicated object takes care of the component integration.

In contrast, other methods break up existing relationships between the composed components, and perform a recomposition by rebinding (Tosic et al., 2002). Typical ways of doing this include the following:

1. Replacing one component at a time, such as replacing the old component with a new component. In this case, the composition structure is relatively static.
2. Breaking the composition structure by replacing two or more components at a time.

A Model for Web Service Integration

Issues for Service Partner Integration

Web service technologies provide middleware to facilitate interoperability, just like CORBA and COM support the interoperability between the traditional components in a component-based distributed system (CBDS). However, interoperability between Web service components is more difficult because interaction between Web service components is driven by business logic, which is affected by the situation of the business market or business objectives, and thus changes from time to time. This is reflected not only in the dynamism of connections between Web service components but also in component behaviour.

Dynamism

A feasible model for integrating service partners must allow for the evolution of Web service components by providing loose coupling. It should minimize the impact of changes in behaviour and interface of Web service components; alterations made to one or more Web service components should have little effect on the overall service. It is also desirable that the model should be able to detect changes in Web service component interfaces.

Dealing with Semantic Differences

Usually, each company has its own understanding of business logic. This may result in different interpretations of the same data that was exchanged between companies. A successful model for integrating Web service components should also take this into consideration. There has to be a mechanism to filter out such differences during the process of aggregating heterogeneous Web service components.

Handling Service Capability Differences

The capabilities of Web service components offered by different business entities can be different. For instance, one Web service component belonging to a travel agent provides online air ticket booking with Qantas, while another provides a similar service with Virgin Blue. The booking systems can offer different services, for example one providing information on waiting lists of flights, the other simply reporting that a particular booking is on waiting list. A successful model also has to adopt a mechanism to handle such issues. Availability of information about the service capabilities of aggregated Web service components can help the model to efficiently select the most suitable component according to the end user's requirement. In this sense, the overhead of building connections between Web service components can be dramatically reduced.

Quality of Service (QoS)

In practice, the Internet may not be as stable as we expect. Due to a variety of reasons, such as virus attack, network crash, bandwidth limitations, and so forth, some aggregated Web service components may not be accessed for a certain period of time, or may disappear permanently. A good model for integrating Web service components should cater to these problems, and, for example, provide a method to select alternative components. To make sure the quality of the value-added service is maintained, the model has to be able to detect the failure of certain Web service components.

Overview of the Proposed Model

As shown in Figure 3, the proposed model employes a layered architecture to integrate the service partners. The model separates the task of integration into two subtasks: integrating semantically equivalent service components in the service collector layer, and integrating semantically non-equivalent service components in the composite service collector layer.

The bottom layer consists a pool of heterogeneous Web service components that are distributed on the Internet.

The service collectors provide a homogenized interface, and thereby uniform access to several, semantically equivalent components. That is, the interface heterogeneity of service components is handled in the service collector layer; an interface conversion takes place there. All mapping and other information needed for the conversion is restricted to this layer, enabling the layers above to be independent of the actual implementations. This layer also contains relevant mechanisms to handle the key issues discussed in the previous section.

Figure 3. Overview of the proposed model

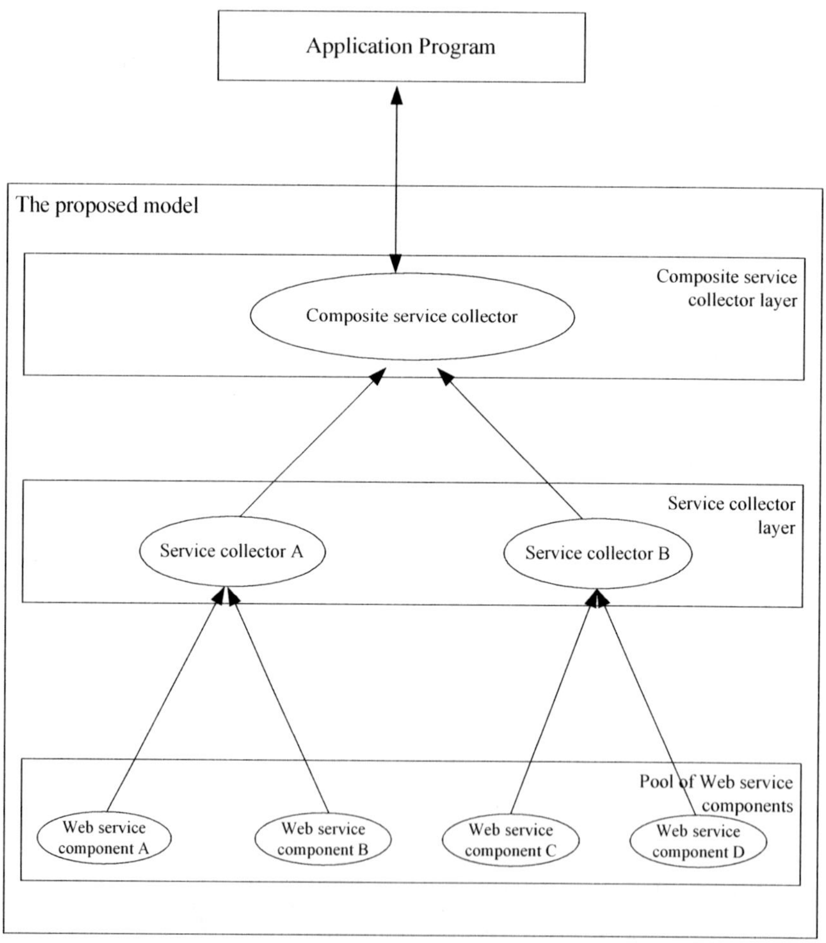

Application programs interact with the composite service collector layer, which provides the value-added service composed of services offered by the service-collector layer. This layer can include transactional information related to the invocation of multiple Web service components.

The interface conversion process spreading across the different layers is depicted in Figure 4.

The Service Collector Layer

A service collector pulls together Web service components that are semantically equivalent, alternative implementations of a particular function, and each collector offers a semantically different service. When invoked, a service collector delegates an incoming request to a Web service component. Each collector performs interface conversion; that is, builds a joint, homogenized interface by mapping the interface of each component into the common interface.

To help portability, each service collector is implemented as a Web service component, and contains the following modules: service collector coordinator, mapping file(s) and a description file as shown in Figure 5.

Figure 4. Building a general interface

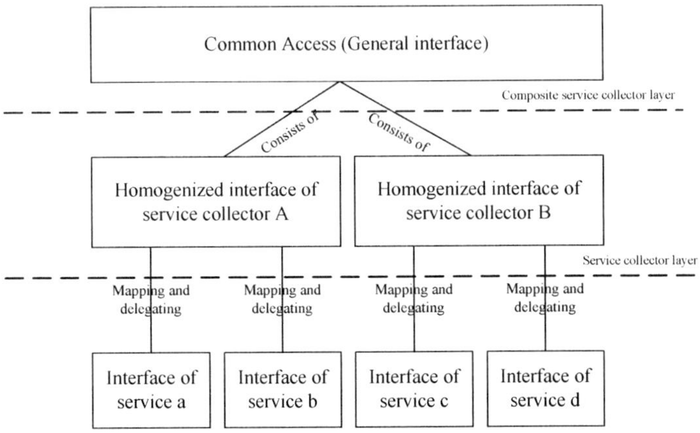

Figure 5. Modules of a service collector

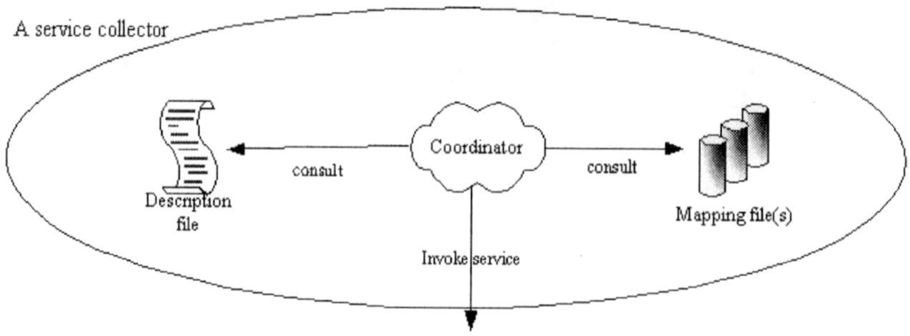

The Service Collector Coordinator

A service collector coordinator has the following major responsibilities:

- Communicating with the upper layer, accepting requests from there and sending the results back.
- Selecting the most suitable component to serve the request, based on the content of the request and the description file of the aggregated components. The actual selection mechanism depends on the implementer's requirements, such as performance, bandwidth, and reliability.
- Mapping the interface of the service collector to the interface of the selected Web service component. This involves mapping of input parameters, output parameters, and mapping the operations. Parameter mapping can be one to one or one to many, depending on the specific situation.
- Communicating with the lower layer. This primarily means invoking the selected component's operation and accepting the return data from the operation. The relevant source code is of the following form:

```
ServiceCollector.operation (input parameter1, input parameter2) {
        SelectedComponent.operation (parameter1, parameter2) {
        .......
        .......
```

```
            return output;
        }
        return ServiceCollector.output = SelectedComponent.output;
    }
```

Service messages, such as start and stop service component, are also exchanged with the lower layer.

- Communicating with the upper layer. The composite service collector calls the different service collectors according to the content of the client request, and the invocation of other service collectors may depend on the result of invoking the current service collector.
- Communicating with UDDI to retrieve the WSDL interface of the selected Web service component. It should be noted that aggregated Web service components may be inaccessible for some reasons, such as a network problem, or an amendment may have been made to the WSDL interface. Consequently, every time after the coordinator selects a Web service component to process a request, it should retrieve the WSDL interface of the selected component through the UDDI service in order to ensure availability and consistency. To address QoS, if the selected component does not respond within a specific period of time, the coordinator selects an alternative component to process the request.

Mapping File(s)

Each aggregated component has a corresponding mapping file that has the same name as the component, with the extension ".properties" appended. The mapping files are created by the implementer manually and are called by the service collector coordinator dynamically. When the service collector coordinator selects a specific Web service component, it loads the relevant mapping file. To extract data from the XML mapping files, a java class implements a Java XML parser. An excerpt from an XML mapping file is given as follows.

```
<interface_mapping>
    <operation_mapping>
        <source name="ServiceCollector">selectCourse</source>
```

```
        <target name="AggregatedComponent">courseSelect</target>
    </operation_mapping>
    <parameter_mapping>
        <source type="input">courseCode</source>
        <target type="input">courseName</target>
    </parameter_mapping>
    <parameter_mapping>
        <source type="input">studentNo </source>
        <target type="input">studentNo</target>
    </parameter_mapping>
    <return_mapping direction="targetToSource">
        <source type="output">confirmation </source>
        <target type="output">confirmation</target>
    </return_mapping>
</interface_mapping>
```

Description File

This file, also in XML format, contains specifications of the aggregated components, such as service description, name, location, interface data, and rank. The description files serve two purposes. First, they support the selection by providing ranks and service specifications of the aggregated Web service components. Key words, such as "course selection," are used to describe the service and a rank is decided based on the implementer's selection mechanisms. The second purpose is to help the service collector coordinator to check consistency between the retrieved interface and the stored interface of the same service component. If any inconsistencies are detected, the service collector coordinator chooses another Web service component to process the request and returns a warning message. An extract from a description file is shown in the following:

```
<component name="CourseSelector">
    <rank>1</rank>
    <function> course selector</function>
```

```
    <location> url address</location>
    <operation name="courseSelect">
        <input parameter="courseName"/>
        <input parameter="studentNo"/>
        <output parameter="confirmation"/>
    </operation>
</component>
```

To browse the interface data and extract data from this description file, another class implementing the Java XML parser was created.

If a Web service component wants to be integrated in this layered framework, it needs to register with a service collector. This step depends on the commercial contract between the provider of the component and implementer of this framework.

Interface Conversion Issues

WSDL Interface vs. Java Interface

The service collector can have two interfaces: a WSDL interface and a Java interface, or it can have Java interface only. The advantage of the first option is that the service collector can also advertise its service through the UDDI server via the WSDL interface. In a sense, a service collector with a WSDL interface becomes a Web service component, which can be accessed through HTTP. External users can access then the service collector directly, without the composite service collector service layer. In contrast, when having only a Java interface, the only way to access the service collector is via the composite service collector layer. However, from another point of view, this disadvantage can be an advantage as it restricts access; having only a Java interface can prevent unauthorized users from accessing the service collector directly. In addition, having only a Java interface makes the mapping between a Java interface (of the service collector) and WSDL interfaces (of aggregated Web service components) much easier. In our implementation we used only Java interfaces.

Advantages of Interface Conversion

The primary advantage of interface conversion in this layer is that aggregations of semantically equivalent components can be built rapidly, because a service collector does not need to provide any implementation details about the service it offers. In other words, code from the semantically equivalent components does not need to be moved to service collectors. All a service collector needs to do is to build the mapping between its homogenized interface and the interface of aggregated components. Another advantage of this technique is that the description file and mapping file provide a convenient and highly effective way to represent interface data of aggregated Web service components, as well as that of service collectors, in such a way that the coordinator can quickly locate the attributes of a specific interface.

The Composite Service Layer

The composite service layer is built on top of the service collector layer. It provides the end user with general access to a value-added, composite service that aggregates multiple service collectors. Compared with the service collector layer, the aggregation in this layer is more straightforward. The different service collectors provide different services, and the composite service layer extracts interface data from all service collectors to provide a general interface to the value-added service.

In general, there are three kinds of service collector compositions, as listed below. It is the requirements of the implementer that determines which one of them is adopted (Yang, 2003).

- **Sequential Service Composition:** Constituent service collectors are invoked one by one. The invocation of one service collector depends on the output of the result of the preceding service's invocation. The major disadvantage of this composition is that if one service collector crashes, the whole composition will become unavailable.
- **Parallel Service Composition:** In this case, all the constituent service collectors are invoked simultaneously and independently. Such composition, however, may not always be suitable for practical execution, as there are cases when the execution of one service collector is dependent on another service collector.

- **Combined Service Composition:** In this case, sequential service composition and parallel service composition are combined. This is the most realistic scenario, but also requires the most complex approach.

Finally, metadata about all of the aggregated service collectors also needs to be stored. Metadata describes capability, location, and access information of services collectors. In summary, this layer contains the following components: a composite service coordinator, metadata storage, and an execution model file as shown in Figure 6.

The Composite Service Coordinator

Its responsibilities include:

1. Decomposing the requests into several subtasks.
2. Designating each subtask to a service collector for processing. In this sense, the coordinator needs to check the metadata storage.
3. Invoking service collectors by exchanging the information with the service collector coordinator.
4. Synthesizing the results from collectors and returning the synthesized results to the calling application.

Figure 6. A composite service collector

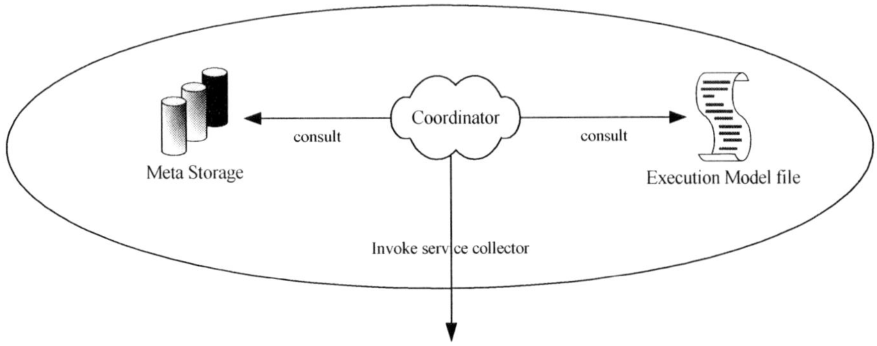

Metadata Storage

This storage file serves the same purpose as the description file residing in the service collector layer. It describes capability, location, and access information of the services collectors. The coordinator uses these metadata to locate, browse and invoke the service collectors.

```
<collector name="Collector_CourseSelector">
   <function> course selector</function>
   <location> url address</location>
   <operation name="courseSelect">
         <input parameter="courseName"/>
         <input parameter="studentNo"/>
        <output parameter="confirmation"/>
    </operation>
</collector>
```

Execution Model File

This file is about how to compose service collectors and is created manually. According to this file, the coordinator invokes the service collectors in the format of sequence, parallel, or combined. It can be also presented in an XML file format.

Implementation of the Proposed Model

Web Service Technologies

Most of the major players have developed Web service technologies: IBM has WebSphere (websphere), Microsoft has published its Web Services Software Development Kit (SDK), and more recently its .NET platform (.NET). Sun Microsystems has developed the Java Web Service Developer Pack (JWSDP). Most features of these products are very similar or the same, and it is the application platform that will eventually decide on the application of a particular product.

Our platform is based on Java, so we chose Sun's Java Web Service Developer Pack 1.2 (JWSDP1.2) for implementation. Several Web service components form a pool, and a Web page using Java Server Pages (JSP) technology enables visits from Web browsers to, for example, allow a human user to access the service. Our implementation has shown that the proposed model is realizable with existing technologies.

In the next section a brief introduction of JWSDP1.2 is given. Then, the overview of the implementation is presented together with the scope of the implementation. Finally, the details of the implementation are described.

Java Web Services Developer Pack (JWSDP1.2)

JWSDP1.2 is an integrated toolkit that allows Java developers to create Java applications as Web service components based on XML, SOAP, and a host of open standards. This toolkit provides a set of tools and application program interfaces (APIs) for creating and deploying Web services and Web service clients; in essence, it is a collection of existing and new Java technology. The toolkit consists of the following major components, as shown in Figure 7.

- Java API for XML-based remote procedure calls (JAX-RPC) to handle the invocation of a method on a remote Web service component.
- Java API for XML Registries (JAXR), to communicate with the UDDI server to publish, access, and query the WSDL interfaces of Web service components.

Figure 7. Java WSDP 1.2

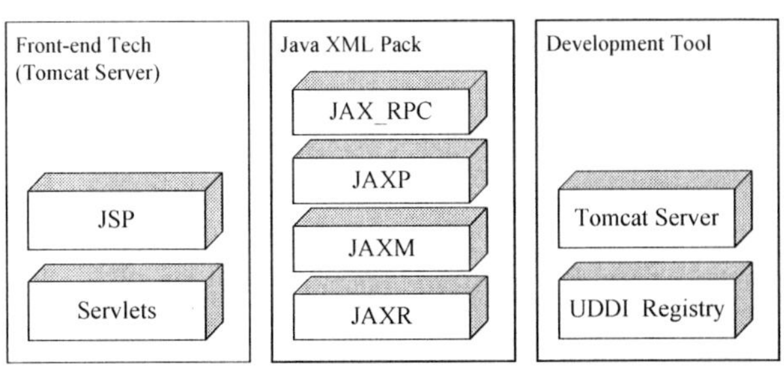

- Java API for XML Messaging (JAXM), APIs for sending and receiving SOAP messages between Web service components.
- Apache Tomcat server, which allows the developers to test their Web service components locally.
- Java WSDP Registry Server (UDDI), an implementation of a "standalone" UDDI 2.0 server.
- Java Server Page, which enables the developer to build a dynamic front-end Web page of a Web service component.

Besides those components, JWSDP also provides several auxiliary tools, which can dramatically increase the efficiency of creating and deploying the Web service components. These include a WSDL stub compiler named "wscompile" that generates a WSDL file for the interface of a Web service component, and a compiler, which generates Java classes from a given WSDL file. In addition, JWSDP comes bundled with Ant, which allows the developer to compile and deploy Web service components on Tomcat at the same time.

Overview of Implementation

Environment of Implementation

The implementation of the proposed model was developed under Linux Redhat 8.0, with JDK1.4.1 and JWSDP1.2 installed. The proposed model was implemented in the Java programming language under JWSDP1.2. Since the focus of this work is handling interface heterogeneity of Web service components, our implementation has concentrated on interface conversion, which is carried out by the service collector layer.

Scope of Implementation

In our experiments, four Web service components (City Campus Bookshop, Bundoora Campus Bookshop, Science Faculty Enrolment, and Business Faculty Enrolment) represented the pool of Web service components. They were Java applications and deployed on a Tomcat server provided by JWSDP.

In the service collector layer, we implemented two service collectors (Bookshop Collector that aggregates a City Campus Bookshop and a Bundoora Campus

Bookshop, and Enrolment Collector that combines Science Faculty Enrolment and Business Faculty Enrolment). Each service collector had a coordinator and two mapping files. However, we did not use any description files that are the basis of a selection policy, as they are determined by the enterprise's business policy, which is not what we want to examine here.

We implemented a simple service coordinator and sequential composition of the service collectors was adopted; that is, the invocation of the service collectors residing in the service collector layer is sequential. This coordinator, accessed through a browser, invoked the service collectors on behalf of the end user and forwarded the result from lower layers to the end user, as shown in Figure 8.

The user interface of the system is shown in Figure 9.

Figure 8. Implementation of the proposed model

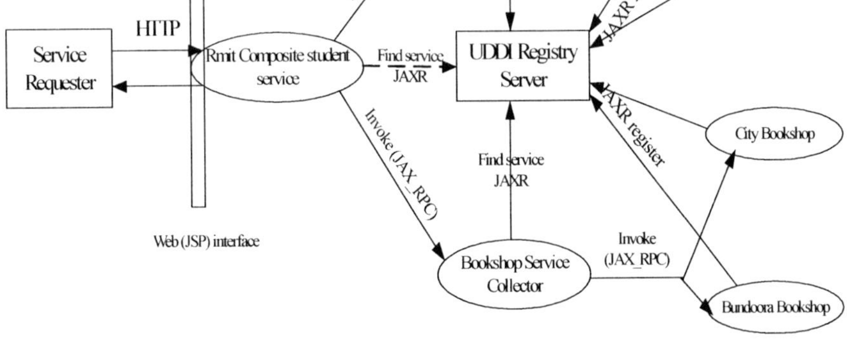

Figure 9. The user interface of the composite student service (CSS)

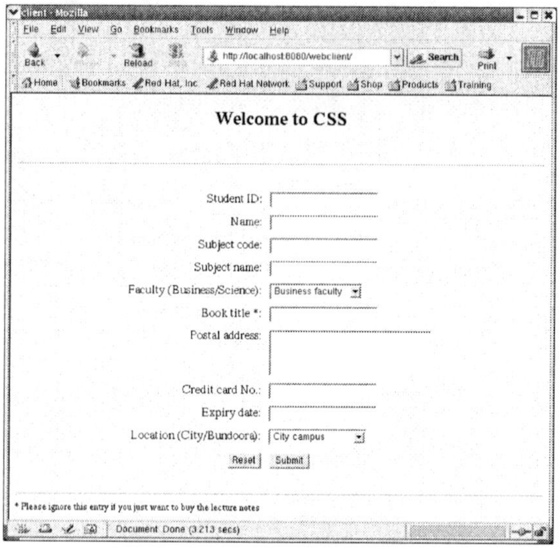

Discussion

Applicability to Smart Organizations

The proposed solution not only assists in communication between different entities, but also allows a wide range of Web services to cooperate and perform tasks together. Smart organization can utilize these features to improve their *internetworking abilities*. At the same time, adding, replacing, and removing Web services becomes simple with the proposed solution, and this facilitates smart organizations' *adaptation to new environments*. A straightforward implementation, as discussed below, can ensure smooth functioning in a dynamic environment.

Implementation Issues

The implementation shows that our proposed solution is realizable. Interface conversion is a stable and efficient method to build connections between components in an ad hoc environment. The layered structure adopted by the proposed model handles interface heterogeneity well. With this model, the task of providing a composite service is decomposed into two subtasks. The first is creating a service collector layer, which homogenizes the semantically equivalent Web service components. The second subtask is aggregating the different service collectors; that is, composing a value-added service from the individual services offered by the service collectors.

Through the adoption of mapping files and selection policy files, the coordinators of the service collectors can dynamically select a Web service component and build the composite service. We have seen that interface conversion and dynamic mapping between the interfaces and is easy to implement. The proposed solution has highlighted the five issues described in earlier sections: compatibility with current XML-based standards, supporting the property of dynamism of Web service components, dealing with semantic differences, handling service capability differences, and the issue of the quality of service. In the following we look at these issues in detail.

Compatibility with Current XML-Based Standards

The proposed model is compatible with current XML-based technology. In our solution, each Web service component has a WSDL interface, which is accessible from the Web and published on the UDDI registry server. A requester can issue a query with the UDDI registry server for the location and description of the requested Web service component, in our implementation by using JAXR APIs. Then it can invoke the Web service component, for example by using JAX_RPC APIs.

The middle layer does not know which Web service components will be selected until runtime, since it has to dynamically make the selection based on the selection policy and the data input by the end user. As a consequence, a dynamic invocation interface (DII) had to be adopted, the service collector acting as a DII client. The invocation between the top layer and the middle layer is different, since the location and interface of the service collectors are relatively static. Therefore, each service collector has a stub object, which is generated by "wscompile" from the WSDL interface of the service collector and resides in the top layer. This local stub object acts as a proxy for the service collector (shown in Figure 10). In our implementation, a dynamic proxy client was created for the top layer to interact with the middle layer.

Figure 10. Invocation between three layers of the proposed model

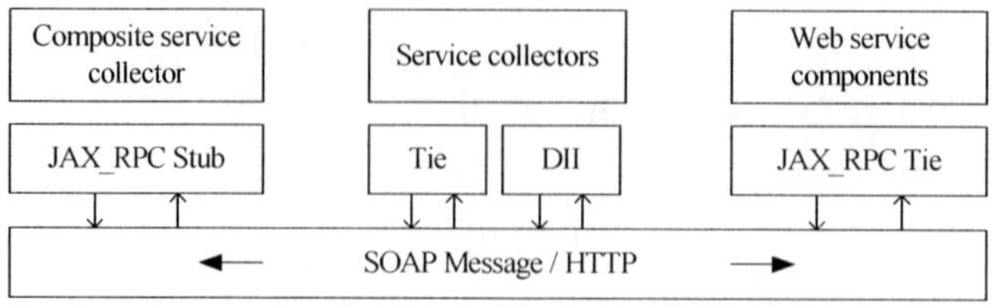

Supporting Dynamism of Web Service Components

In the proposed model, a change made to the interface and location of one Web service component does not have any significant impact on the integration of service partners; all we have to do is to make a minor amendment to the mapping file corresponding to that particular Web service component.

Dealing with Semantic Differences

Interface conversion is an efficient and straightforward method for handling semantic differences that may appear in the way in which different Web service components understand the business logic. In our example, when enrolling in a subject, one faculty expects the students to provide a subject code, while another faculty may instead expect the students to present a subject title. In the proposed model, the interface conversion consisting of parameter mapping, method mapping and return value mapping, helps to resolve such conflicts. With the mapping files residing in the middle and top layers, the semantic differences between interfaces of different Web service components can be dynamically filtered out.

Handling Service Capability Differences

Besides semantic differences, the proposed model also takes capability differences into consideration. Web service components can be different in their service capabilities. The description file in the middle layer and the metastorage file in the top layer provide the necessary information about service capabilities of aggregated Web service components. With these two files acting as selection policies, the coordinators of the top and middle layers know which component's service should be selected and invoked to answer a user request. For example, a decision can be based on the content of "location" field provided by the user: If "City" or "Bundoora" is inputted, the City Bookshop service or the Bundoora Bookshop service will be invoked, respectively.

Ensuring Quality of Service

To ensure Quality of Service (QoS), the proposed model has remedial solutions that come into play when some aggregated Web service components fail. When integrating service partners, the performance and availability of integration depends on the performance and availability of the elementary Web service components. A major advantage of the proposed solution is that the Web service components are loosely coupled, and the failure of a Web service component does not prevent the invocation of other Web service components; if one component becomes unavailable, the service collector can choose another one to process the request. For example, if the City Bookshop service does not respond within a certain time, the Bookshop Collector would invoke the Bundoora Bookshop service. In other words, the proposed model ensures quality of service by supporting the on-site replacement of a Web service component with a similar one.

Limitations of the Proposed Solution

Although the proposed solution satisfactorily addresses the above issues, it still has some limitations. Interface conversion is a good way to address semantic differences and service capability differences; however, it is heavily dependent on mapping files that are created manually by the programmer. Automating that process is still a challenge.

Another issue is that the service collector itself can become a bottleneck and single point of failure. In the proposed model, different service collectors provide different services and there are no alternatives to replace a failed service collector.

Conclusion

Knowledge sharing and efficient use of expertise are major issues for organizations that want to operate the smart way. Sharing skills with others within the organization, making them available externally, or using external resources not only reduce redundancies but also enable combined, value-added services to be offered. Collaboration and cooperation have to be managed cleverly, however, and sound infrastructure support is needed.

The Internet and the World Wide Web has proven to be an efficient communication medium, and traditional work management systems such as workflows are finding their ways to utilize this platform efficiently. In addition to content providing, the Web is also used as a service provider, and Web services are becoming increasingly popular.

We have seen that Web services can be combined very efficiently to build value-added services. However, an important problem with Web service integration is that services are heterogeneous in many aspects, such as having different interfaces, using different technologies, and so forth. This chapter described a method of dealing with heterogeneity by employing a layered architecture for integration. In one layer, a composite service collector combines different services in a sequential, parallel or hybrid way to make a value-added service, where each service represents a step in a process, similarly to traditional workflows.

The proposed method can also handle alternatives for a particular task, and select one of them for integration. Service collectors in the layer below convert the interface of selected components to the interface expected by the composite service collectors.

Implementation proved the method is feasible and can be used in real-life environments.

References

Amazon (2004). Retrieved March 15, 2005, from http://www.amazon.com/gp/browse.html/103-7642094-8511817?%5Fencoding=UTF8&node=3435361

Benatallah, B., Dumas, M., Sheng, Q. Z., & Ngu, A. H. H. (2002). Declarative composition and peer-to-peer provisioning of dynamic Web services. In *Proceedings of the 18th International Conference on Data Engineering* (pp. 297-308). IEEE.

Benatallah, B., Sheng, Q. Z., & Dumas, M. (2003). The self-serve environment for Web services composition. *IEEE Internet Computing, 7*(1), 40-48.

Booth, D., Haas, H., McCabe, F., Newcomer, E., Champion, M., Ferris, C., et al. (2004). *Web services architecture*. Retrieved March 15, 2005, from http://www.w3.org/TR/2004/NOTE-ws-arch-20040211/

Cerami, E. (2002) *Web services essentials*. O'Reilly & Associates.

Christensen, E., Curbera, F., Meredith, G., & Weerawarana, S. (2001). *Web services description language (WSDL) 1.1*. Retrieved March 15, 2005, from http://www.w3.org/TR/wsdl

Clark, M., Fletcher, P., Hanson, J. J., Irani, R., Waterhouse, M., & Thelin, J. (2002). *Web service business strategies and architectures*. Expert Press.

Della Torre Cicalese, C., & Rotenstreich, S. (1999). Behavioral specification of distributed software component interfaces. *Computer,* 32(7) 46-53.

de Figureiredo Pires, P. (2002). *Webtransact: A framework for specifying and coordinating reliable Web services compositions* (Technical Report ES-578/02 PESC/Coppe Federal University of Rio de Janeiro). Retrieved March 15, 2005, from http://www.cos.ufrj.br/~pires/webTransact.pdf

Filos, E., & Banahan, E. (2000). Will the organisation disappear? The challenges of the new economy and future perspectives. In L. M. Camarinha-Matos, H. Afsarmanesh, & R. J. Rabelo (Eds.), *E-business & virtual enterprises* (pp. 3-20). Dordrecht: Kluwer.

Gudgin, M., Hadley, M., Mendelsohn, N., Moreau, J., & Nielsen, H. F. (2003). *SOAP version 1.2 part 1: Messaging framework*. Retrieved March 15, 2005, from http://www.w3.org/TR/soap12-part1/

JWSDP. Retrieved March 15, 2005, from http://java.sun.com/webservices/jwsdp/index.jsp

mindreef (2004). Retrieved March 15, 2005, from http://www.mindreef.com/

netegrity (2004). Retrieved March 15, 2005, from http://www.netegrity.com/products/

.NET. Retrieved March 15, 2005, from http://www.microsoft.com/net/

Parasoft (2004). Retrieved March 15, 2005, from http://www.parasoft.com/

Perkins, D. (2002). *King Arthur's round table: How collaborative conversations create smart organisations*. John Wiley & Sons.

SDK. Retrieved March 15, 2005, from http://msdn.microsoft.com/workshop/management/tools/wpsdk.asp

Snell, J. M. (2004, October). *Resource-oriented vs. activity-oriented Web services*. Retrieved March 15, 2005, from http://www-106.ibm.com/developerworks/webservices/library/ws-restvsoap/

stylusstudio (2004). Retrieved March 15, 2005, from http://www.stylusstudio.com/

Tilley, S., Gerdes, J., Hamilton, T., Huang, S., Muller, H., & Wong, K. (2002). Adoption challenges in migrating to Web services. In *Proceedings of the 4th International Workshop on Web Site Evolution (WSE'02)*, Montréal, Canada (pp. 21-29).

Tosic, V., Pagurek, B., Esfandiari, B., & Patel, K. (2002). *On various approaches to dynamic adaptation of distributed component compositions* (Tech. Rep. OCIECE-02-02). Ottawa-Carleton Institute for Electrical and Computer Engineering (OCIECE).

UDDI.org (2000). *UDDI Technical White Paper*. Retrieved March 15, 2005, from http://www.uddi.org/pubs/Iru_UDDI_Technical_White_Paper.pdf

websphere. Retrieved March 15, 2005, from http://www-306.ibm.com/software/websphere/

Yang, J. (2003). Web service componentization. *Communications of the ACM, 46*(10), 35-40.

Yang, J., Papazoglou, M. P., & van den Heuvel, W. (2002). Tackling the challenges of service composition in e-marketplaces. In *Proceedings of 12th International Workshop on Research Issues in Data Engineering: Engineering E-Commerce/E-Business Systems (RIDE-2EC 2002)*, San Jose (pp. 125-133).

Chapter IX

Grid Technology for Smart Organizations

Gergely Sipos, Hungarian Academy of Sciences, Hungary

Péter Kacsuk, Hungarian Academy of Sciences, Hungary

Abstract

This chapter summarizes the most relevant results that grid research achieved in the last decade, it presents the actual issues of the topic, and it outlines how current and future results from this area can contribute to smart organizations. At the first place the basic goal of the Grid is presented and its state-of-the-art, service-based realization is discussed. This global infrastructure will one day connect together diverse types of hardware and software elements, abstracting them out as intelligent autonomous agents that can discover and collaborate with each other on demand. The middle part of the chapter introduces two potential middleware technologies that service grids can be built on. They are the Web services-based open grid services architecture (OGSA) and Jini. The final part of the chapter presents the future of service grids and the important role these flexible infrastructures will probably have in the life of smart organizations.

Introduction

Since the late 1990s—more or less since the idea of "The Grid" first came up—virtual organizations were always "the dream to realize" for the grid research community. In the grid computing sense, a virtual organization connects geographically and architecturally dispersed resources together in order to make them accessible for a well-defined group of people in some convenient way (Foster, Kesselman, & Tuecke, 2001). These people are usually referred to as "e-scientists," because they require virtual organizations to solve computationally intensive (sometimes grand challenge) problems they identified in theoretical or practical science (Foster & Kesselman, 1999). Grid-based virtual organizations are the tools for them to demolish the walls built by limited capacities of computational power, storage space, and network bandwidth. If grid research manages to create virtual organizations on the top of existing computer networks—primarily on the top of the Internet—then these e-scientists will be able to access remote resources whenever and wherever they would like to.

In the next section the motivations of grid computing and the necessity of a global Grid infrastructure are presented. The chapter introduces the service-oriented architecture, the approach that is believed would one day realize this vision of the global infrastructure. The chapter reviews the general structure of service-oriented grids and the building blocks they are consisting of: the grid services. The aim is to demonstrate that service-oriented grids can act as underlying layers for highly intelligent giant distributed systems.

Because the "service-oriented architecture" is a system design principle and not an implementation, service grids have to choose a technology that brings the idea into practice. The chapter presents the two most potential candidates, the technologies that were more or less involved in almost every grid project in the last few years. These technologies are the Web services extended with the open grid services architecture (OGSA) and Jini. The chapter presents how OGSA—with the additional support of the open grid services infrastructure (OGSI) and the Web services resources framework (WSRF)—can turn the Web services technology into a "grid middleware." The chapter points out why Jini—the middleware always criticized for its Java dependency—can result in more intelligent and more convenient grid services than OGSA.

Finally, the last part of the chapter demonstrates how smart organizations could use grid-based virtual organizations as flexible infrastructures to understand

and influence the environment they are operating in, to adapt themselves to changing situations, and to efficiently overcome their all-time problems.

Introduction to the Grid

Each day more and more complex equations and mathematical models are formulized within different fields of science. Although computers and computer networks are said to be the most dynamically developing tools of our age, they are always few steps behind theorems: They never have enough capacity to solve any kind of computational task within the preferred periods. While in the last few decades larger and larger supercomputers and computer clusters tried to cope with the computationally intensive jobs, in the future the global Grid will provide a better solution. The "grid-approach" is based on the aggregation of resources that may be situated far away from each other. In contrast with supercomputers and clusters, virtual computers of the Grid will consist of hundreds, thousands, or even more relatively small-capacity machines and network components in order to generate the required capacities. Connecting more and more elements into these virtual computers, sooner or later the size of any supercomputer or cluster can be overgrown. If grid computing succeeds in forming "virtual supercomputers" from individual machines, then the grid approach will definitely win over supercomputing in the long run.

There are millions and millions of machines—PCs, clusters, mainframes, or supercomputers—connected to the Internet, to the global computer network. Quite usually these machines, although they are turned on, do not perform any computation. Typically this is the situation at companies where the employees use the IT infrastructure during the day but the machines are not turned off for the nights. The wasted computational power gave the inspiration for the pioneers of grid computing to develop high-level infrastructures that can optimize the utilization of computational and storage resources (Litzkow, Livny, & Mutka, 1988). The key is the sharing of these resources among the members of a well-defined community in a controlled and secure way. This resource-sharing does not necessarily end up with allowing others to access the obvious "computer" or "storage" functionalities of the connected machines. Other functionalities can be shared in a similar way as well. Such functionality can be, for example, the printing facility of an Internet-enabled office printer.

Basically, any kind of function that some integrated device or connected periphery realizes can be shared in this way.

When one day the Grid will offer this broad spectrum of hardware devices, users will not have to invest into new, sometimes very expensive hardware elements when they are required. Remote instances will be available through the Grid. Of course, it does not mean that in the future only the providers of the grid-infrastructure will buy hardware. First of all, some kind of computers or computer-like mobile devices will be necessary to access the Grid. Second, sooner or later the Grid will operate on a financial basis, which means that resource owners will charge users in some way. Thus, before one will decide whether to buy or rather "borrow" a specific device through the Grid, one will estimate which option provides better value for the same price.

Shortly summarized, the aim of grid computing is to realize resource sharing, to make networked elements accessible for remote users in a secure way. To achieve this goal, a new software layer has to be built on the top of the Internet. This layer hides the low-level network protocols and standards and turns the Internet into a global pool of resources, or rather pool of functionalities the interconnected resources provide. If one requires some kind of networked resource for a given purpose, this software layer should "route" the request to a suitable device in a transparent way. If researchers develop this layer, usually called the "grid middleware", then the Grid becomes a reality.

In the early years of grid computing, the Grid has been usually referred to as the analogy of the electric power grid in the information age (Foster & Kasselman, 1999) (That is where the name "Grid" came from.) While devices can use the electric grid to get the electricity they require to function, computer programs will use the Grid to get processors, memories, and other devices for similar purposes. The analogy is very emphatic, since end devices do not have to take care about the several components the electric grid consist of, they just have to be plugged into a socket and they work. They do not know where the power comes from, the several elements the network consists of, or the way in which these elements produce electricity. The same should be true for the Grid as well. Computer programs should not have to take care of the machines that provide computational power, storage capacity, or other specialized functions. Once they are plugged into the Grid they can get what they need on demand.

Although the comparison suggests that a similar network can be developed for computer programs than the one that has been established for electrical devices, the situation is not so simple. The power grid provides only one thing, electricity, in one well-known "format" (frequency and voltage values)[1].

Consequently, electric devices do not have to specify the "functionality" they need when they are plugged in. In contrast, the Grid will contain various different provider resources and all of the functions they offer must individually appear for the clients. Accordingly, clients of the grid middleware will have to specify what functionalities they need at the time they are plugged-in. This negotiation is totally missing from the electric power grid.

There is another significant difference between the two networks. While electricity—the "functionality" that electric grids provide—can be separated from the plants, physically transmitted through wires into sockets and from there into end devices, the Grid behaves differently. Computational capacity, storage capacity, and all the other functions cannot be separated from the computers and transmitted into sockets. Since the Grid will be an additional layer on the top of the current computer networks, in the Grid not the presented functionalities, rather just information that represents these functionalities will flow.

The Service-Oriented Grid Approach

In the previous section the Grid, as a global pool of functionalities provided by computers and their peripheries, has been introduced. Computer programs can and will access the Grid to use these computational resources on demand. While this statement was absolutely true in the early years of grid computing (called the resource-oriented Grid generation), today it has to be completed. While the original goal of grid research was to realize the sharing of computational resources, it was found that in reality not the resources but functionalities these hardware elements, together with the pre-installed software are able to provide, have to be shared. It results that not only low-level hardware capacities, but also higher-level, aggregated functionalities can also appear in the Grid. The term "service" can be defined in several different ways. In the most generic sense it refers to an entity (hardware and/or software) that provides some functionality to others. Consequently, the Grid will be the global pool of services instead of resources, and the grid middleware has to help find services, and not resources.

This service-centered approach is more general than the resource based one. Grid services can embody the previously discussed low-level functionalities (e.g., "computation" or "storage") beside high-level ones such as "SMS-sending" or "matrix inversion." While the plain program execution facility can be provided by a single processor, SMS-sending obviously assumes intelligent cooperation from several hardware and software elements. While low-level grid services will more or less exactly represent physical resources, high-level grid services will embody virtual resources that contain not only hardware capacity but built-in intelligence, as well.

The service-based Grid will realize resource sharing, the original goal of grid computing. Because every service will directly or indirectly embody physical resources, clients will exploit hardware capacities when they utilize services. Moreover, if a service provider uses other grid services to provide some high-level functionality for its clients, these clients will harness the capacity of all the low-level machines when they use the high-level service. The schema of such a multi-level service usage can be seen in Figure 1.

In the upper part of Figure 1, the client of a high-level grid service can be seen. This high-level service is actually the client of other (lower-level) Grid services. Since every service encapsulates one or more hardware element(s), the

Figure 1. Aggregation of resources in the service Grid

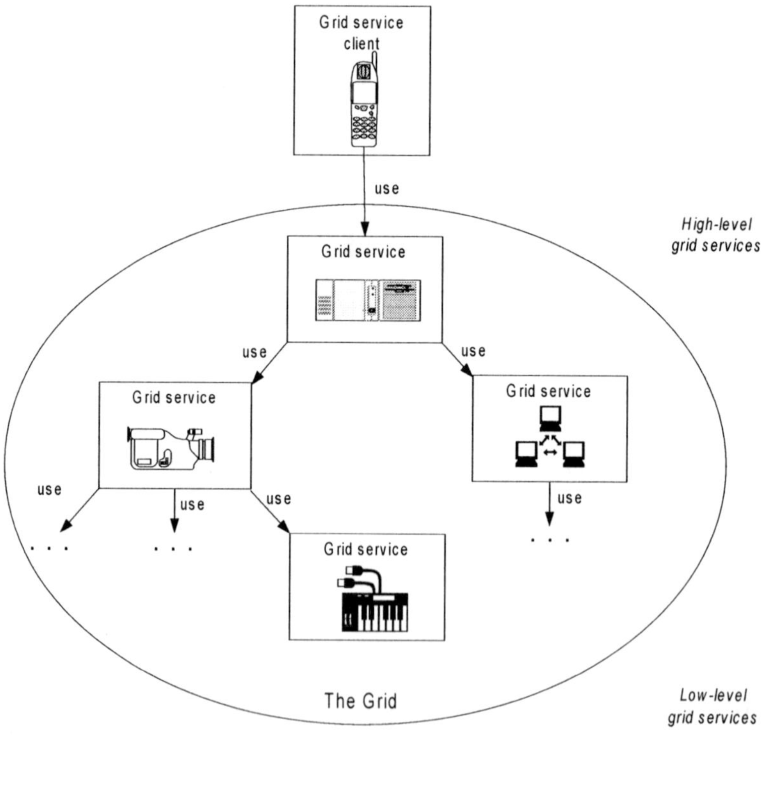

capacity of all these elements will be allocated in some way for the end-client of the "service-graph." Notice that the higher level service an end-client uses, and the more services are participating at each hierarchical level, the more resources are being aggregated. In other words, the wider and the deeper a service-graph is, the more hardware capacity and software intelligence it represents. Consequently, the usage of large service graphs will be the key for researchers who need enormous computational and storage power.

The Architecture of the Service-Oriented Grid

Participants of a service-oriented system can be classified into two basic categories: service providers and service consumers. In case of the Grid the classification is not so simple. A machine that appears as a grid service provider can also use other grid services.

It has been mentioned in the grid introductory section that grid clients have to specify the properties of the services they require when they access the middleware. "The Grid" has to reply to such a request with a list of providers that meet the requirements. The client can choose the best provider from the received list, and using the meta-data being associated with it, it can begin the usage of the service. As it can be felt from this scenario, the Grid is not only a set of hardware and software resources wrapped into services and offered for others. It has a "default functionality" as well. This functionality is some kind of service repository that clients (and high-level providers) can use to find grid services. This repository has to operate even if there are no "real" services in the Grid at all, thus it can be regarded as a service that the grid middleware itself provides. The name of this repository is grid information system, and it offers service registry and service lookup facilities for grid participants and grid clients. It is obvious that this information system must be implemented in a fault-tolerant and scalable way. What is much more important to emphasize is that the grid information system must provide its service for computer programs and not for humans. While its fault-tolerant and scalable properties assure that "the Grid layer" will operate even in heavy load, the last one is unavoidable to make the multi-layered aggregation of services possible (see Figure 1).

The "developed for computer programs and not for humans" mentality will be relevant not only to the information system of the Grid, but also to every service it contains. If it is not realized then the Grid will become a single-layered service architecture just like the World Wide Web is today. The "providers" of the

Web—Web servers, CGI scripts, servlet containers, the servlets they manage, and so forth—are not able to discover and use back-end services on demand. (Since all of these components generate replies to requests they can be regarded as service providers.) The Grid will be different. Grid service programs—and through these programs the resources they represent—will be able to automatically join together, to form complex tools. These complex tools will appear as high-level "superservices" for the end-users. Only a machine-usable information system can guarantee that high-level service providers will be able to find lower-level providers automatically. Since the whole discovery process can happen in an automated way, complex service-graphs can be shaped without manual administration, without additional help. The central role of the grid information system is depicted in Figure 2, where the registration, discovery, and usage processes of a grid service are also presented.

When a resource would like to join the Grid its first task is to locate the grid information system (1). After the information system has been located, the resource publishes its service by sending some description about it to the remote registry (2). This description has to give enough information for its future clients (remember, they are computer programs) about the properties of service and the way it can be used. When a client (which can be integrated in a higher-level service provider) appears in the network to find a grid service for a specific purpose, first it has to find the information system, as well (3). The client then sends a description about the required service (4) and the information system returns a list with the suitable providers (5). The client program chooses the best record from the list and begins the usage of the service it characterizes (6). The provider, while it serves the client, can find lower-level

Figure 2. The registration, discovery and usage scenario of a grid service

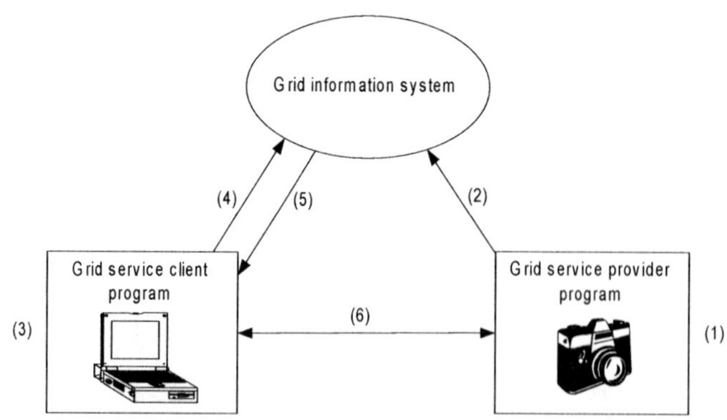

services similarly to how it has been discovered earlier and can integrate them as building blocks into its own service. The result is a multilayered "service tree" similar to the one that has been presented in Figure 1. The end-user—the human who controls the client program—perceives only the aggregated functionality, without knowing or taking care of the distributed way it is actually provided.

The previous two sections presented the goal of grid computing, the resource and then the latest, service-oriented grid approach. At the end, the inner structure of the service-based Grid has been presented. Although it seems very probable today that the information system-based solution will give the core of the Grid, there are still several unanswered questions as to how it should work. How can a worldwide and fault-tolerant registry be created that always up to date? What would be the most suitable way and protocols for the client programs to access it? How to develop a grid service program that makes the functionality of some hardware or software resource accessible remotely? What about client programs that can act on the behalf of their owners or hosting machines? How can these services and clients understand each other automatically?

We have several different answers for each question. Unfortunately, none of them solve all of our problems in practice. The issues are far more complex than to solve them in a general way. Extra assumptions must be added to the previously presented service-oriented architecture in order to get "something work". An implementation of this architecture is required, which meets the requirements of at least a subset of the potential users. The next sections introduce two different implementations that could be suitable for this purpose. They are OGSA and Jini. Although both technologies add different extra assumptions to the general service-oriented vision, they both can be used to solve the "Grid problem," known as "flexible, secure, coordinated resource sharing among dynamic collections of individuals, institutions, and resources what we refer to as virtual organizations" (Foster, Kesselman, & Tuecke, 2001).

The Web Services Technology

In 2003 a paper entitled "The Physiology of the Grid" was published (Foster & Kesselman, 2003). It presented a conception about a Grid that uses the Web services implementation of the service-oriented architecture (Glass, 2002) in order to form virtual organizations for distributed computing purposes. The paper actually proposed patterns to standardize the behavior of participants of Web services-based networks. These patterns are called open grid services architecture (OGSA). Since that time, no widely usable OGSA Grid has appeared on the top of Web services. One of its reasons is probably the continuously developing Web services standards. Because of this, the section tries to present only the general ideas behind the Web services technology, but not its latest specification.

It was mentioned in the previous section that the Web provides services for humans, and not for computer programs. When an end-user accesses the Web, he or she controls the providers—such as CGI scripts, Web servers, servlets—through client-side entities like HTML pages or applets (all hosted by the client-side browser). Although these provider programs can access physical or logical resources automatically, this is not the way in which grid services should use each other. Even if a CGI program—that serves for example a Web-shop—is connected to another service provider, usually a database, it does not find this resource on demand. The system administrator has to define in advance which database and how the CGI program has to contact. When the program later receives a client request—for example, the content of a form that specifies an order—the program sends the information into this preconfigured database. The data is then usually read from the database by an administrator, who then contacts an appropriate factory or warehouse and accomplishes the order manually. It can be stated that Web-based systems cannot process such orders automatically. This is because the Web provides information for humans and not for programs, so the components of the above scenario cannot discover partners automatically.

The Web services technology would like to make a difference here. Using the core elements of the Web—servers and clients that use standardized protocols and markup languages to communicate with each other—the Web services technology introduced two standards: Web service description language (WSDL) and simple object access protocol (SOAP). The core Web protocols, together with these standards, can generate an infrastructure for intelligent devices, to enable them the automated integration of functionalities through the Internet.

WSDL and SOAP

With the WSDL standard Web service provider programs can create a machine-processable description about the functionalities they offer. While today the "interface" of a simple Web-enabled service (for example a Web-shop or an SMS-sender) is an HTML page that can be understood only by humans, in the future, based on WSDL, intelligent programs can also become service "users."

Before a Web service client program can access a provider, it has to obtain the WSDL document that belongs to its service. Web service clients can download WSDL documents from Web servers just like browsers download HTML pages today (WS Activities, 2002). It follows that Web service providers have to make their WSDL documents accessible, just as if they were ordinary HTML pages.

After a suitably implemented client program downloads the WSDL description of a Web service, it can parse the document to generate a stub, which is needed to access the remote server. The task of this stub is to turn the client requests into the server-specific format and to send it "through the wire." On the server side, a so-called skeleton component receives the request, calls the real provider program (which performs the requested action) and sends a result back on the same route. This scenario can be seen in Figure 3.

Although the presented request (1), action (2), answer (3) procedure-triplet seems to be the same that happens when a browser downloads a Web page, there are two important differences:

1. In the Web services world the main purpose of this three-step procedure is not to download or upload of some information from or to the server, but to initiate the server-side action. The information transfer here is "only"

Figure 3. Usage of a Web service

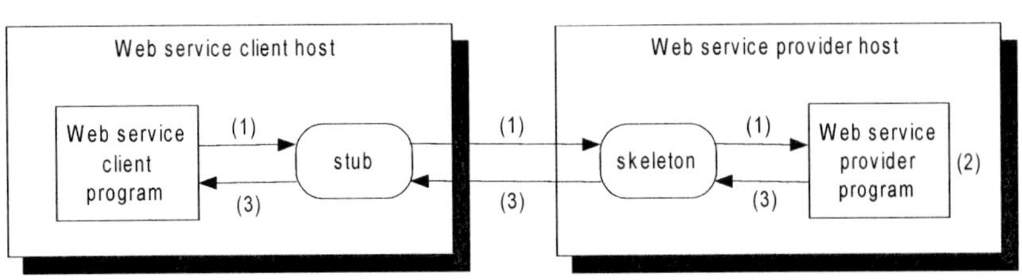

a side effect. While in the case of the Web this server-side action is at most some manipulation on a database or file system, in the Web services world diverse, even physical actions can be imagined. A Web service that acts as the back-end engine of a Web-shop can control production lines and server robots directly or indirectly through additional software layers. In this case, the generated server-side action contains physical elements besides the obvious information parsing process.

2. A Web service provider program can discover and use other Web services on demand, so multi-layered, client-server architectures can come into being. Using only Web technologies, this process cannot be carried out.

As it has been stated above, WSDL files are treated as ordinary Web pages during network transmission. It follows that they are sent over HTTP from the providers to their clients. The Web services technology uses the HTTP protocol for every other communicational purpose too, so service request and response messages are HTTP request and responses.

To enable automated interaction between networked resources, the Web services technology introduced a standard format for service requests and reply messages. This specification is called SOAP. Both WSDL and SOAP achieve the machine-processable format by building onto the Extended Markup Language (XML) standard (W3C, XML). XML is a text-based, semi-structured data representation language developed to share data between computers in a platform and programming language neutral way. Because Web service providers and clients communicate with each other using XML messages transmitted over HTTP, the Web services standard could remain a totally platform- and language- independent solution. It means that the provider and client programs can be implemented in any programming language, and they can be executed on any hardware and software platform.

In this section, the core Web services standard has been discussed. It has been shown how WSDL and SOAP were built onto HTTP and XML in order to support the automated collaboration of machines. Once a Web service client has obtained a WSDL document, it can contact the remote server and it can initiate any actions the remote server supports. Although Web service clients could obtain WSDL references from their end-users just like browsers are given with URL addresses today, obviously this is not the solution the Grid requires. Some kind of registry should be created where Web services can register themselves and client applications can use to discover providers automatically.

A critical point is to decide what to store in these Web service registries. Links (URLs) to WSDL documents are essential. What else is required? Giving a general answer to this question is not easy, and probably impossible. A description that sufficiently characterizes a Web service for one specific client may give ambiguous or imperfect information for another. Since Web service provider programs can represent broad spectrum of functionalities, the creation of a general "registry standard" is impossible. Nevertheless, creating registries for similar types of services is possible. Such a solution is the Universal Description, Discovery, and Integration (UDDI) registry (UDDI standard).

UDDI Registries

The aim of the UDDI registries is to enable the integration of business processes through the Internet. Companies can advertise and make their "real" services accessible as Web services for their customers with UDDI. If a company advertises itself with a traditional Web page, then only humans can understand what the company is specializing in. Finding current offers on the Web is usually time-consuming and sometimes a very involved process.

To eliminate these and other similar problems, companies can start Web services that turn information about their profiles into machine-processable documents. These "business" Web services can be published in UDDI registries. To make the developers" lives easier, UDDI registries are implemented as Web services, so client applications that already know the Web Services protocols are able to reach these metadirectories "by default."

The functionality that a UDDI registry offers is actually a "notice board" that companies can use to advertise themselves for client programs. So, while the Web can be used to reach customers directly, UDDI can be applied to do business in the "service world." If a client application contacts a UDDI host, it can search among the registered services. Once a suitable service is found, its WSDL file can be downloaded and the client can do business with the company through the service that represents it. In this sense, UDDI services are "virtual markets" where companies can meet their customers in order to initiate on-demand business-to-business and business-to-customer transactions.

UDDI registries associate information such as the name of the company the service belongs to, its profile, telephone and fax numbers, its industry code, location, and so on with each registered WSDL link. Based on this metadata, the clients can find companies for various purposes. A Web service client can search for office equipment vendors in Houston to find out which one of them can deliver 50 computer desks within the shortest period of time. Another client can search for his well-known industrial partner "Harbour Metal, Inc." to get current offers on compound steel. The possible variations for the scenarios are limited only by the intelligence built into the provider and client programs.

The notion of the UDDI community today is that a global UDDI-network will be operated. Companies that intend to do e-business in the introduced way can register themselves at this global information system. The UDDI hosts of this network will be in contact with each other, and they will synchronize the contents of their databases on a regular basis. Thus, once a company registers its Web service into any of these UDDI hosts, it can be sure that sooner or later the registration will appear in all the others, as well. Consequently, there will be no need to contact every UDDI host to find the most appropriate company for a particular purpose; sending the "service lookup" request to only one registry should be enough. Figure 4 depicts the idea of this global UDDI network.

Figure 4. Global UDDI network

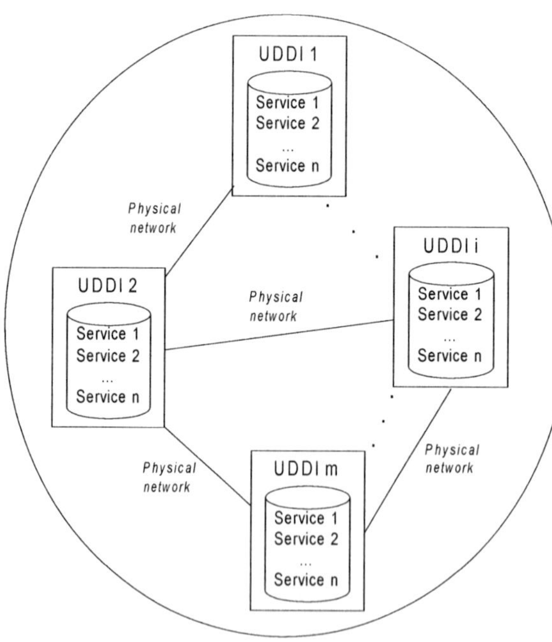

Although the approach is promising in the sense that neither companies nor customers have to contact every UDDI host when they register or search for services, it poses several practical issues. What kind of UDDI implementation would be able to cope with the huge mass of data? Is the global replication really necessary? If it is not, what kind of data distribution should be applied? Organize UDDI hosts into hierarchical layers or allocate a single-layered structure? Although big IT companies are already operating UDDI services that synchronize their databases on a regular basis, they still do not have enough registered records and clients to feel the negative effects of the global replication. It has to be emphasized that data distribution and replication issues can be one point where the basically "client-server" model of the service-oriented architecture can fail, and where the Grid has to use some peer-to-peer solutions instead (Fox et al., 2003).

As the summary of the section, Figure 5 presents the full registration, discovery, and usage scenario of a Web service through a UDDI registry.

First the Web service provider program downloads the WSDL document of the UDDI registry from a known URL (1). This URL can be preconfigured or can be defined by the user of the provider program manually. Based on the downloaded description file the provider generates a stub (2) and then transfers

Figure 5. UDDI based registration, discovery and usage procedures of a Web service

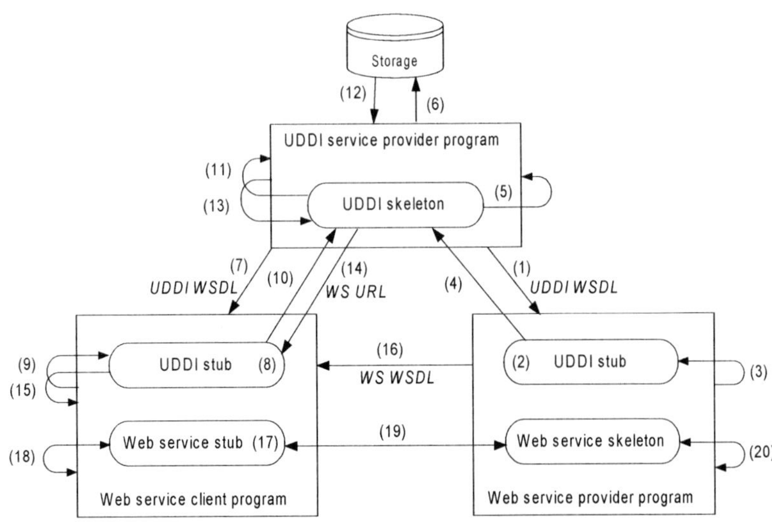

the service description in a SOAP format (3-4-5). The UDDI provider saves it into some permanent storage and waits for the client (6).

After the client application is started, its first task is also to download the WSDL document of the UDDI registry (7). It generates a stub (8) and sends an XML formatted service lookup request to the UDDI server (9-10-11). The UDDI provider performs a query on its database (12), then sends the result back as a SOAP response message to the client (13-14-15). The client chooses the most suitable provider from the list, and from the WSDL URL contained by the chosen record downloads the WSDL document of the provider (16). It generates another stub (17) and applies it to access the remote site (18-19-20).

Although both the service provider, the service client, and even the UDDI provider can get information from its local user (or administrator) at any point during the scenario, it is not the recommended way for their operation. The less these programs rely on users the more automated the whole process becomes. Although the Web services architecture is quite popular today, most of the time it is used not because it enables multilevel automated service aggregation, but because it helps to create simple client-server and 3-tier distributed systems in a platform independent way.

OGSA-Enabled Grids

In the previous section the Web services technology, as the bottom layer of OGSA-enabled service-oriented grids, has been introduced. As one implementation of the service-oriented architecture, the Web services standard gives a standard-set for the questions that have been posed at the end of the introductory section. By wrapping computer programs and hardware devices into Web services, one can create an environment in which these entities can automatically understand and use each other. If these programs apply UDDI registries, then they can discover each other on demand. Although it seems that the Web services technology by itself provides solution for the Grid community, there are two reasons why this statement is false:

1. The Web services technology specifies the way in which stateless functionalities can be expressed as services, while the Grid would like to turn mostly stateful resources into providers.

2. The Web services technology supposes static environments with stable, long-living providers. In the Grid, providers will come and go unceasingly.

To overcome the limitations of the Web services technology, caused by the inadequate assumptions it has on network participants, in 2003 Foster and Kesselman proposed OGSA. OGSA is basically a set of principles that prescribe the behavior of Web services. If one implements a Web service that follows the OGSA principles, the result can be called an OGSA grid service. If a distributed system contains only OGSA grid services, then it can be called an OGSA grid. If it overgrows the boundary of an intranet and enables access for clients and providers from all over the world, a global OGSA Grid comes into existence.

Representing Stateful Resources

As it could have been seen in the previous section, the Web services technology exploits the widely accepted Web protocols to give an implementation for the service-oriented architecture. Service interfaces (WSDL files) are accessible for clients just like HTML documents are published today; moreover, Web services clients and providers communicate with each other using similar messages that Web servers and browsers are using today. Disregarding the fact that Web services messages contain XML and not HTML code, they are equal and they are transmitted over HTTP. One important consequence of the stateless HTTP protocol that Web services provide stateless function just like Web servers do. The statelessness of a service means that the provider server supposes that incoming requests are independent from each other. If they are independent, then no information about the incoming requests has to be stored at the server side, the server program does not keep count a state, and it remains stateless.

Of course, because of a stateless transmission protocol, the high-level Web service does not have to be necessarily stateless. If server programs apply stateful back-end resources to keep count of their states, then even stateful services can be developed on the top of any stateless protocol. This is the way CGI programs, servlets, and other similar solutions turn Web servers into stateful providers. Web service developers can do exactly the same. They could use databases and other stateful resources to implement stateful services. The point is that the Web services technology does not specify a standard way for it. Web service developers have to find the most suitable solution every time when they want to create stateful providers.

Although stateless servers were sufficient for the original purpose of the Web—namely, to publish static documents to remote clients—and they are enough to create basic Web services, they are absolutely insufficient for the Grid. As mentioned earlier, the goal of grid computing is to share resources among network partners (the members of a virtual organization) in a secure and controlled way. The service-oriented approach could be a good solution to achieve it: If we wrap resources into service programs users can access them remotely, and other providers can use them as basic building blocks to generate more complex functionalities.

Because resources represent stateful functionalities, grid computing has to find some general way in which these stateful elements can be turned to services. Although current Web technologies were taken into consideration, for different reasons none of them suits the needs of the Grid. The perfect solution should support grid developers to create stateful services quickly and easily. To standardize the creation and usage of stateful Web services, OGSA introduced the following three concepts:

1. Grid Service Handler (GSH) and Grid Service Reference (GSR)
2. Service factory
3. Service Data Element (SDE)

Identifying things is necessary, whatever field of life is being examined. In computer science, operating systems identify users and processes, database managers identify tables and records, object-oriented programming languages identify existing objects. There will be no difference in the Grid either: Grid services have to be identified. GSHs were introduced exactly for this purpose. According to the OGSA principles, every grid service is to be identified with a GSH. Providers and clients can use these GSHs to refer to services when they register or search for them. GSHs are globally unique and a given GSH belongs to the same grid service through its lifetime.

Besides a GSH, every OGSA Grid service has to have a GSR. Unlike GSHs, GSRs are not necessarily unique, and a grid service can have different GSRs in the different periods of its life. While the GSH identifies the grid service, its GSR describes the way it can be actually used. A GSR specifies important things like the interface of the server or metadescriptions that characterize it. Because OGSA supposes that in the future highly intelligent grid services, which can implement different interfaces in the different phases of their life, can exist, it separated service behaviors from service identities. While the identity (GSH) of a service cannot change, its GSR can be changed at any time.

Since an OGSA-enabled grid service is actually a Web service that is implemented in a predefined way, the introduction of the GSH and GSR does not bring any novelty into the Web services world. The WSDL document of a service can be regarded as its GSR, and the URL this document can be downloaded from as its GSH. Since in the Web services world a WSDL file specifies the public features of a service, it exactly meets the requirements of an OGSA GSR. When a Web service changes its behavior, it simply modifies its WSDL (its GSR), but does not modify the URL that points to its location (its GSH). It is important to emphasize that although in the Web service world WSDLs and URLs play the roles of OGSA, GSRs and GSHs, OGSA does not adhere to WSDL documents as GSRs and URLs as GSHs. With these technology independent naming convention OGSA probably wanted to keep itself separated from the all-time Web services standards and wanted to let the door open for future changes.

The second element that helps developers to build stateful Web services is the notion of service factories. Service factories are special Web services that can be used to create stateful service instances. As it has been described in the Web services section, services can be started before their clients. In OGSA grids this is not always the case. Although stateless grid services operate exactly in this way, stateful grid services are created explicitly for the requests of clients. Stateful OGSA services (programs that make stateful resource[s] accessible) are started by grid service factories, thus OGSA clients can get references to stateful services from service factories instead of UDDI registries. Consequently, a resource owner who would like to share a resource in an OGSA-based virtual organization has to start not the service that makes the stateful resource accessible, but rather just a service that can start such types of instances when the clients need them. Although the introduction of service factories does not seem to make much sense, the aim of OGSA was to standardize the way in which resource capacities are allocated for users.

Stateful services are usually not "exclusive gateways" to the physical resources; they usually share a common resource with other similar instances. If a service factory starts multiple stateful service instances on the same resource, then the capacity of the physical device or the software has to be divided among these

Figure 6. Multiple stateful grid service instances sharing a single resource

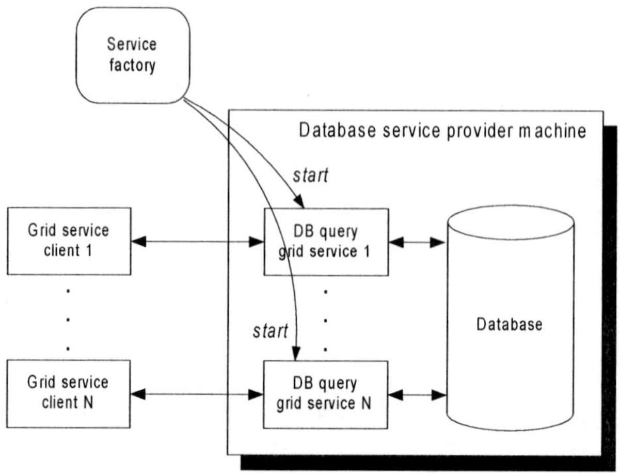

instances. In case of a stateful grid service that provides access to a database, for example, it is more logical to start a new service instance for every client on the same physical machine, then to copy the full database onto an idle host. As a result, multiple clients will manipulate on the same database—on the same stateful resource—but each of them will do it through its own stateful service instance. This example is presented in Figure 6.

The aim of OGSA with service factories was to virtualize resources, to make clients feel that they all have their own resources. The idea is not new—OGSA only ordered the developers to use this schema every time they need stateful grid services.

Although OGSA introduces the service factory notion, it does not specify how these factories should be implemented. While one factory can generate threads as the different stateful service instances, another one can use independent operation system processes for the same purpose. Developers have to estimate the most suitable solutions for their all-time needs.

There is another important issue here that OGSA does not deal with: "What should be returned to the client as the result of a service instantiation request?" The GSH of the new instance? Its GSR? Or something else? OGSA left this question unanswered to let developers choose the most suitable implementations for their current needs.

The next extension that OGSA gave to the Web services standard is the Service Data Element (SDE) concept. Since stateful services act on the behalf of their back-end resources, a stateful service can perform a client-requested operation only if the actual state of its back-end resource enables it. A developer can choose basically from two options to implement this decision-making procedure:

1. Develop a service program that queries the state of its back-end resource every time a client request arrives.
2. Develop a service program that stores the state of the resource in the memory as some kind of data fields and updates these values regularly or on demand. In this case, the service can decide whether or not to perform a client-requested operation without contacting the back-end resource.

While programs that apply the first approach always make the right decision, they sometimes have long response times, because the back-end communication usually takes much longer than to evaluate the values of some field variables. This is especially true if the back-end resource is a different physical device from the one the Web service program is running on.

With the SDE concept, OGSA supports the development of the second type of services. A service data element denotes a data field inside a stateful Web service that stores one attribute of a back-end resource. By storing SDEs inside the providers programs, Web services can more –or less keep up with the states of the resources they maintain. The more SDEs a Web service stores and the more often their values are updated, the more accurately the server knows its environment. If back-end resources can notify their wrapper services every time they change their states, SDEs can perfectly express the inner states of physical devices. Figure 7 presents the connection between a back-end resource and its corresponding SDEs.

The patterns that OGSA introduced to make Web services suitable for grid computing purposes are not new. Service identification, resource virtualization, and data fields were already used both in the Web and in the Web services world before OGSA. Web and Web services developers already applied similar solutions when they created stateful services. The point is that they did not do it in a standard way. OGSA collected the best practices, gave exact definitions for the applied methods, and presented them for the grid community.

Figure 7. Expressing the state of a resource with OGSA SDEs

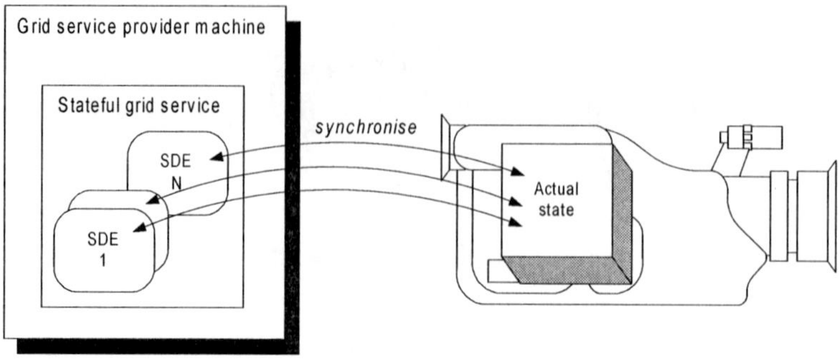

Handling Dynamic Behavior

The main aim of the developers of the Web services technology was to support the integration of business processes through the Internet. In reality, business environments are quite static. Companies do not disappear from the market from one second to the other, so Web services that represent them in the virtual world do not really have to take care of dynamism. The notion of the Web services community was that Web service provider machines would operate all around the world for months, or even years, statically. This idea is based on the procedures that companies apply today when they install their IT infrastructures. There are, among others, dedicated Web servers, mail servers, database servers, and application servers, and all of them administrate their tasks for long periods. Since these providers operate continuously, their clients do not have to take care when they were started and when they would stop functioning.

In the Grid this assumption will not be true. Since computers, tools, and equipments can join and leave the Grid at any time, thousands and thousands of servers will appear and disappear every second. A computer that is idle because its owner went to have a coffee can join the Grid to offer its computational power. A taxi, being equipped with an onboard computer, can join to announce that it welcomes passengers. A finishing machine in a repair workshop can join to let its server robot know it is empty, so the next work piece can be carried. These were only a few of the possible scenarios from the life of the Grid, but we hope they help one imagine how dynamic this environment would be. Because the Web services standard does not specify patterns to handle this level of dynamism, OGSA introduced three concepts:

1. Service lifetime management
2. SDE notification
3. HandleMap services

While the previously discussed GSH-GRS, service factory, and SDE concepts regulated the creation of stateful Web services, the enumerated patterns give support for the management and destruction of such stateful components. The service lifetime management helps providers handle the problem of disappearing clients. HandleMap services help clients adapt themselves to new or changing providers. The SDE notification framework defines solutions for asynchronous server-to-client communication.

As discussed in the previous section, stateful OGSA grid service instances are created expressly for the requests of the clients. The result of such an instantiation request is some reference to the new stateful service instance. As it has been presented in Figure 6, such an instance does not necessarily guarantee exclusive access to the resource. The client usually gets only a "slice" of the capacity the hosting device is able to provide. Whatever the situation is, a stateful instance logically or physically allocates some part of the back-end resource for its client. The approach poses an important question: "When can this resource part be freed?" The answer is obvious: The part has to be freed when its client does not need it any more. This is the case when the client permanently disappears from the network, when its user closes the application or when it collapses due to some failure. Since the freeing procedure of a resource practically happens when the client of a stateful service disappears, the question can be rephrased: "How can a stateful service know when its client disappears from the network?" If the client gives the resource up voluntarily, then it can notify its provider. The stateful instance can deallocate all the resources that have been allocated for the client so far, and it can self-destruct. There are two problems with this solution: First, it assumes the existence of a "destructor" method in the interface of every stateful service. Second, it does not give relief for the problem of involuntarily disappearing clients. OGSA introduced the service lifetime concept to give a general, fault-tolerant solution for every situation.

In OGSA grids every stateful service has to have a lifetime. This lifetime denotes the length of the time the service instance lives for. Every service instance has to maintain its lifetime and when the period elapses the instance has to free up every resource it has allocated and has to self-destruct. Stateful services can get their original lifetime values from their factories during their instantiation processes. Factories can choose these initial lifetime values arbitrarily, or they can allow clients to give preferences on them. (In the latter case clients have to send the preferred lifetime values, as arguments of the instantiation requests.) If at any time after the instantiation process the client realizes that the granted lifetime value is too short, it can set a new lifetime for the service, but only if the stateful instance supports it. If factories set relatively short initial lifetime values for stateful instances and let rather the clients control these values on demand, then resources can be freed as early as possible.

It can be stated here again that OGSA "reinvented the wheel." On the Web, several sites apply similar functionality to optimize resource allocation. Portals that offer free e-mail addresses delete mailboxes if the owners do not log in within a given period of time. In this case, mailboxes are the stateful instances (they allocate storage capacity), they have initial lifetimes of a few weeks, and these values are automatically extended with every log in. OGSA declared that the same approach is an adequate solution in the Grid, as well.

The next OGSA concept is SDE notification. SDE notification is the tool for providers to let the clients know about important events that happen inside the shared resources. When a client accesses a stateful service, an event is generated. The server program reacts to this event: It manipulates on the back-end resource, then returns an answer. Other kinds of events can be imagined besides such client-generated events: events that are derived from the back-end resource or the computer that hosts the service provider program. Since a Web service stores the actual state of its environment in SDEs, a server-side event causes the change of at least one SDE. If a service client is interested in some server-side event, it is actually interested in the changing of some SDE. Consequently, stateful providers have to forward SDE values to their clients to notify them about important events. This process is called SDE notification. As it has been stated several times, an OGSA grid service is actually a restricted Web service. The OGSA "rules" does not affect the core Web services protocols, so every OGSA grid service has to use XML messages to communicate with its clients. As a consequence, SDE notification messages have to be XML documents as well.

Although the advantage of state-change notification is indisputable, it makes the lives of clients more complicated. If a client application wants to receive SDE notification messages from a provider, it has to become a server. Because SDE notification is physically an HTTP request sent from a service provider to a service client, some server program that can handle the request must be operating on the client side. This server program must be able to accept the asynchronous incoming calls, to forward them to appropriate handlers, and to return HTTP responses. One of the reasons for the success of the Web is the simple way it can be used as a client. Whatever advantages the SDE notification concept can add to the Grid, the difficulties it causes to clients can question its success in the practice.

The last OGSA extension is the HandleMap service, or HandleMap functionality. This concept denotes a special type of service that supports GSR changes. It was mentioned earlier that OGSA does not specify what the result of a service instantiation request has to be. Obviously, the GSH of the new instance has to appear in the message. If the GSR is also included, the client can generate the stub and use the service at once. What happens if this response does not contain the GSR? The client has to get the GSR by any means, since it is required to generate the communication stub.

HandleMap is the service that helps clients resolve GSHs with valid GSRs. The task of HandleMap providers is to reply for incoming GSHs with appropriate GSRs. With HandleMap services, the clients of a grid can get the GSRs of the stateful grid service instances any time. Consequently, service factories have to register the GSH-GSR pairs of the stateful instances they create before they send GSHs back to their clients.

The necessity of HandleMap services in grids where service instantiation responses do not contain GSRs is obvious. There is another situation in which a HandleMap provider is indispensable: in grids that contain services that change their GSRs during their lifetime. It has been discussed that while a GSH belongs to a service forever, its GSR can be changed any time. If a service changes its GSR then its clients cannot use the stubs that have been generated from the old GSR. Clients have to use HandleMap providers in such a situation to resolve the GSH of the service (it remained the same) with the current GSR. Once the new GSR has been obtained new stubs can be generated and the upgraded service implementation can be accessed. Obviously, these HandleMap providers can return the new GSR for the clients only if the service registers it at once when it is changed. Additionally, grid service providers can use the SDE notification framework to let clients know about such a GSR change event.

The role of HandleMap providers is known. Only one question remained open: "How can a client know which HandleMap provider to use in order to resolve a service GSH?" There could be several possible answers given to this question: If only one HandleMap provider is allowed to operate in a grid, then by necessity, every provider and client will use the same HandleMap registry. If there is real-time data replication between HandleMap providers (similarly to UDDI network has been presented in Figure 4), then it does not matter which HandleMap service a provider or a client uses. A third solution can be an intelligent HandleMap network that is able to route GSH resolving requests among its participants. In this case, a client can contact any provider of this network since the routing layer sends the request to the host where the requested GSR is stored.

OGSA proposes a fourth approach. The idea is based on the introduction of "home HandleMap." Home HandleMap is an attribute that every grid service must have. (It can be regarded as a compulsory SDE). The value of such an attribute is a reference to the HandleMap provider that stores the GSH and the GSR of the service. If a client would like to resolve the GSH of a service with the appropriate GSR, then it has to know the value of the home HandleMap attribute of the service. Once the value is obtained, the client can contact the HandleMap provider and can get the needed GSR. To guarantee that clients would always be able to obtain home HandleMap values, OGSA suggest wrapping these values into the GSHs. Since clients know the GSHs, they can always find the adequate HandleMap providers.

The Present and Future of OGSA: OGSI and WSRF

Although the Web services technology in itself cannot be used as the middleware layer of the service-oriented Grid, its platform and language independent property can be a good basis to develop this global infrastructure. Accordingly, OGSA examined the Web services technology and identified the points where it does not meet the requirements of the grid community. It has been found that the Web services standard does not support the creation and management of dynamic and stateful services. As a result, OGSA introduced the previously discussed service reference, service factory, SDEs usage and notification, lifetime management and HandleMap service concepts in order to turn the Web services infrastructure into a grid middleware.

Although these concepts can probably fill in the gaps, OGSA has not specified how these extensions must be implemented. The described tools are high-level patterns, and not precise specifications a programmer could work with. Take a look, for example, at the GSH concept. It has been described that a GSH is a globally unique ID, that every grid service has exactly one GSH and every GSH has to contain a reference to a home HandleMap provider. This information is enough for a developer to create a GSH implementation, but it is not a specification. If we let two developers create their own GSH implementations, there is little chance that these GSHs will be compatible with each other. This difference is enough to prevent two grids from being interconnected. So, as a conclusion, it can be stated that OGSA gives only a high-level vision about the way Web services-based environments could act as grids, but it cannot achieve the main goal, to develop standard infrastructure that anyone can join, leave, and use on demand.

The document that gives a detailed specification on the top of OGSA is called open grid services infrastructure (OGSI) (Tuecke, Czajkowski, Foster, Frey, Graham, & Kesselman, 2002). If Web services developers apply the detailed OGSI specification besides the high-level OGSA description, then they can produce grid systems that are compatible with each other and that can be interconnected any time. Since in OGSA grids every physical and logical resource is represented by a Web service, OGSI had to give a specification that is consistent with the pure Web services model. Just to recap: A Web service is a software service that accepts requests and produces replies in SOAP format through its well-known interface being described with a WSDL document. The point is that a Web service, and consequently an OGSA grid service as well, keeps contact with its front-end environment through XML messages. (It can use any other protocol to communicate with the back-end resources.)

Since Web service clients can see only XML-based interfaces and messages from the providers, OGSI defined every OGSA-introduced entity in XML. OGSI identified the interfaces and messages that must participate in an OGSA grid, and expressed them in XML. For example, an OGSI-enabled GSHs looks like this:

```
<xsd:element name="handle" type="ogsi:HandleType"/>
<xsd:simpleType name="HandleType">
<xsd:restriction base="xsd:anyURI"/>
</xsd:simpleType>
```

OGSI introduced similar XML description for every previously discussed entity and functionality. Because OGSI clients and providers see each other only through these XML documents, in an OGSI Grid every service and client can run on any computing platform and can be implemented in any programming language.

The extensions that OGSI added to the Web services technology met with a warm response not only within the grid, but also within the Web services community (IBM, 2004). The Web services community realized the necessity of a generic standard that deals with the issues of stateful and dynamic providers, so they integrated the OGSA/OGSI principles into the Web services patterns. Because the Web services community had a few complaints with the OGSI standard, they slightly revised it. The result is a set of six specifications that altogether can be regarded as the latest specification for OGSA (Czajkowski et al., 2004). The name of this standard set is Web Service Resource Framework (WSRF) and it covers the following topics:

1. Resource properties
2. Resource lifetimes
3. Resource addressing
4. Service groups
5. Fault management
6. Notification

As it can be guessed more or less from the names, each of the enumerated standards deals with one of the OGSA extensions just described. Although there are still only a few companies behind the WSRF standard and behind the idea to integrate it into the core Web services protocol stack, the process started and shows a clear trend: Grid computing and Web services technologies are getting closer and closer to each other. First grid computing applied the Web services implementation of the service-oriented architecture to make resource sharing and integration possible. Now, the Web services world builds onto the results of the grid community. If this trend continues maybe one day one of these two areas will incorporate the other.

The Jini Technology

In the first part of the chapter, the Grid, as a global service pool has been presented. The elements of this pool are implemented on the top of the grid middleware, a layer that contains the grid information system and specifies behavioral patterns and protocols. Independently from the technology this middleware uses, its task is to support resources that want to join as service providers and applications that want to discover services. OGSA proposes using the Web services technology for this purpose. In OGSA grids, clients and providers access each other with standardized XML messages, so OGSA components can be developed in any programming language. Although this property makes OGSA grids very attractive, Web services are not the only technology that the grid middleware can use.

This section introduces Jini, another service-oriented technology that can be a potential candidate for the role of the grid middleware. At the time when Jini was introduced, its developers did not have the aim to create virtual organizations. They wanted to enable the automatic discovery, usage, and management of networked devices (Waldo, 1999). As described in the introductory section, this process-triplet paves the way that leads to virtual organizations. As a consequence, grid-based virtual organizations could be defined on the top of either Jini or Web services networks.

The goal of Jini is very similar to the goal of Web services: Give an implementation for the service-oriented architecture. Although resources appear as services both in Web services and in Jini networks, these two types of services are slightly different from each other. While a Web service uses XML documents to communicate with its front-end environment, a Jini service applies Java objects for the same purpose. As a consequence, every Jini service and Jini client program has to be implemented in Java, and every device that would like to participate in a Jini-based distributed system must host a Java Virtual Machine (JVM). Although these seem to be strict restrictions at the first sight, in actuality more and more ordinary devices fulfill them. Recent cell phones and PDAs are already able to execute Java applications. If current trends continue, in the not too distant future all of our every day devices—televisions, DVD and MP3 players, microwave ovens, refrigerators, cars, elevators, industrial robots, barrows—will all be equipped with Java processing units. If this time comes, the Web services technology will loose its advantage of language independency, since the "language of the network" will be Java anyway.

A Jini-based network—with the Jini terminology a djinn—consists of services and clients. Similarly to an OGSA grid service, a Jini service can embody some physical resource (e.g., a video camera), some logical resource (e.g., a matrix multiplication function), or the mixture of the two. While a service from the first type allows its clients to dynamically discover and remotely control one specific tool, providers of the second and third types can realize intelligent control over devices, software components, and low-level providers.

If a client program wants to utilize a Jini service, first it has to obtain the so-called proxy of the service. Proxies are entities that represent the providers to the client applications. Since Jini services use Java objects to maintain contact with their clients, Jini proxies are Java objects, as well. After a client obtains a proxy object, it can use the service through the interface(s) the proxy implements. In some sense proxies in Jini are equivalent to stubs in Web services (See Figure 3). The important difference is that while a Web services stub is only a gateway between the Web service client application and one dedicated server, a Jini proxy can communicate with any number of servers, and it can perform additional actions at the client side, as well (Edwards & Rodden, 2001).

Just as in the Web services world, three possible actions can take place after a Jini server receives a request: The server performs some stateless computation (e.g., computes the optimum of the function received as the input parameter of the call), it manipulates on a stateful resource (e.g., inserts the input parameter into a database), or some combination of the two. After the server finishes any of these actions it sends back a reply to the proxy. The proxy can use the received value to perform some action locally on the client host, or to generate new requests to the same, or to a different, server. After the proxy receives or computes the final result, it can return it to the client. A possible scenario for the proxy-based usage of a Jini service can be seen in Figure 8.

As clearly seen in the picture, Jini proxies hide all the details of servers and resources from the clients. Although this is not a novelty, since stubs do the same in the Web services world, it has been already mentioned that there are significant differences between a stub and a proxy. While Web service stubs only forward requests and answers between client programs and servers, Jini proxies can contain (not necessarily) built-in logic, and thus they can actively participate in the realization of the services. In extreme situations a service proxy can serve its client without communicating with any server at all. On the other hand, some Jini proxies—just like the one depicted in Figure 8—can

Figure 8. The usage of a Jini service

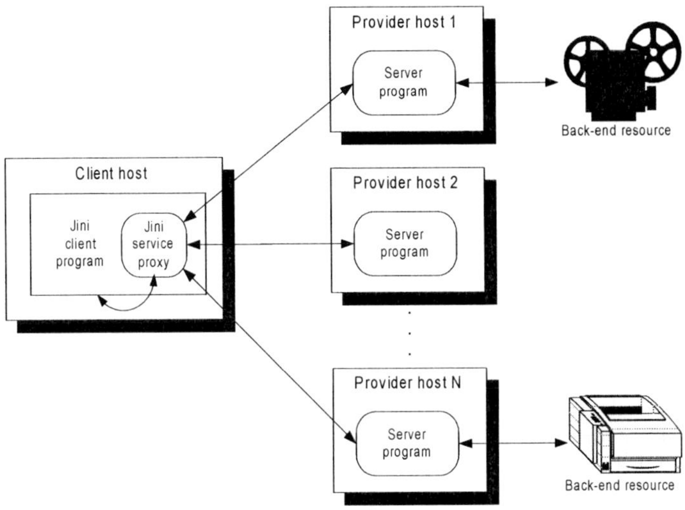

communicate with multiple remote servers to fulfil a single client request. In the Web services world, a stub is unable to contact more than one server.

To demonstrate what these two differences mean in practice, we will see an example service, a "scientific calculator" service that can be used to evaluate mathematical expressions such as "10+34" or "sin(0.34)+(arctg(23))." The "resources" such a service offers are computational capacity and calculation logic. The capacity is necessary to count an expression, while the logic is required to know how to count it (what the different operators mean). First we will see the Web services implementation of the service. First, the interface of the service has to be specified. The future clients of the service will send the formulas for evaluation using this interface. The interface has to be described with WSDL and must be implemented in an optional programming language. After the service is started and discovered by a client, its WSDL document can be used to generate a stub for it. Through the stub the client can send formulas and receive results as presented in Figure 3. We now look at the Jini implementation of the service. There are two possible ways to follow. The first one would give a similar result to the Web services implementation: A server that can be accessed by the clients using a local reference to it, which is in this case not a stub but a proxy. TBoth the proxy and the server must be implemented in Java.

The other solution is much more interesting. Since a Jini proxy is a Java object and every object contains methods, the proxy of the "scientific calculator" can contain the logic required to perform mathematical evaluations. Since every Jini client host is equipped with a processing unit (otherwise they would not be able to execute the client applications) this "intelligent" proxy could perform the evaluation on the client side, without communicating with any remote server. The required capacity is coming from the hosting resource, and the logic is coming from the proxy.

In this example, the drawback of the Web services technology was caused by the fact that client-side service endpoints (the subs) are generated by clients on the basis of a platform-independent document (WSDL file). Since calculation logic (actually, any kind of execution logic) cannot be presented to a client application using a semi-structured document, there is no way to provide any functionality except the message transfer on the client side. Suppose for a minute that the developer of the Web services version of the "scientific calculator" service tries to put the calculation logic into the WSDL document in some way. (Actually, he or she pulls the service interface and the service implementation into the same WSDL document.) Since the stub will be generated on the basis of this document the developer might think: "The intelligent WSDL will result in intelligent stub." While this statement could be even right, there is one huge problem. The client program, which is able to generate an "intelligent stub" from this XML document, has to be so intelligent that the whole stub generation process makes no sense: the created entity will be much "dumber" than the generator client.

In conclusion, we can state that although Jini is not a language-independent solution, the usage of Java bytecode instead of XML has a huge advantage: Real functionality and not only data can flow between clients and providers. We hope that the example of the "scientific calculator" service was enough to demonstrate the point: In case of a Jini service, a proxy, an executable entity, developed by the service provider is transferred to the client host.

A really useful consequence of the usage of mobile code is that Jini proxies can contain user interfaces besides programming interfaces, as well (Edwards & Rodden, 2001). While client applications can utilize proxies through their programming interfaces, humans can achieve the same through the user ones. Figure 9 presents the difference between the two types of interfaces.

If the proxy of a Jini service is equipped with a user interface, then the end-user needs a client—some kind of "Jini service browser"—only to discover the service, but not to actually use it (Juhasz & Sipos, 2004). The service browser

Figure 9. Different types of interfaces can be provided by Jini service proxyies

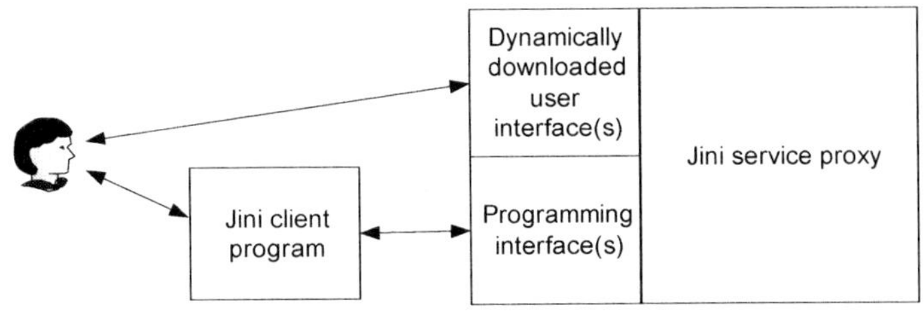

can automatically start the user interface being associated with the proxy and step out of the communication chain. These user interfaces can be consoles, graphical windows, sound recognizers, or basically any solutions that are supported by the client-side input/output devices. Again, if one would like to provide a graphical interface for a Web service by simply extending the WSDL with a description of the user interface, clients would find themselves in the same situation as before: They should have to be much more intelligent than the user interface they generate.

Jini clients, or even the end-users, can utilize services through proxies, and these proxies can provide services in two ways: using their built-in functionalities and calling remote servers. The implementation of a proxy is evident: Since it is a Java object it can contain methods. These methods define the built-in logic and the way in which the proxy has to perform communication with remote servers. Only one question remains open: "What kind of protocol must proxies use to communicate with servers?" The answer for Jini: whatever they wish. Since the back-end communication is hidden from the clients (and from users), proxies can apply any protocol in order to contact remote servers. There is built-in support in Jini (actually in Java) for a wide range of solutions such as Java RMI, CORBA, or XML over HTTP, but those who find them insufficient can use any custom protocol on plain sockets, as well.

Since Jini is an implementation of the service-oriented architecture, it contains an information system, too. The information service in Jini is called lookup service (LUS), and it can be used just like any ordinary Jini service: through its proxy. While providers can access LUSs to advertise their services, client applications can apply them to discover services for their current needs. Because from the users' points of view proxies represent services, the most important purpose of LUSs is to store proxies. Besides proxies, LUSs can store any kind of meta-information—expressed as Java objects—to describe the registered services. These descriptor objects must be generated by the service providers and have to be transmitted to the LUS hosts as part of the proxy registration process. Providers can associate any information with their services in this way. The consequence of this "schemaless" LUS meta-information database structure is that the Jini information system can be used not only for business-related services, but for any kind of service. This is an important difference between UDDI and LUS. Although any kind of meta-information can be attached to a proxy, lookup services have one regulation: A special object that represents a globally unique service ID is compulsory. Such an ID can be regarded as the Jini appropriate of the OGSA GSH. Figure 10 presents all the introduced elements of Jini using a snapshot from the life of a djinn.

Every djinn has to contain at least one LUS. The presented one contains two. These LUSs store the proxies of the available services and the attributes that represent the metadescriptions associated with them. As seen here, the LUSs

Figure 10. The structure of a djinn

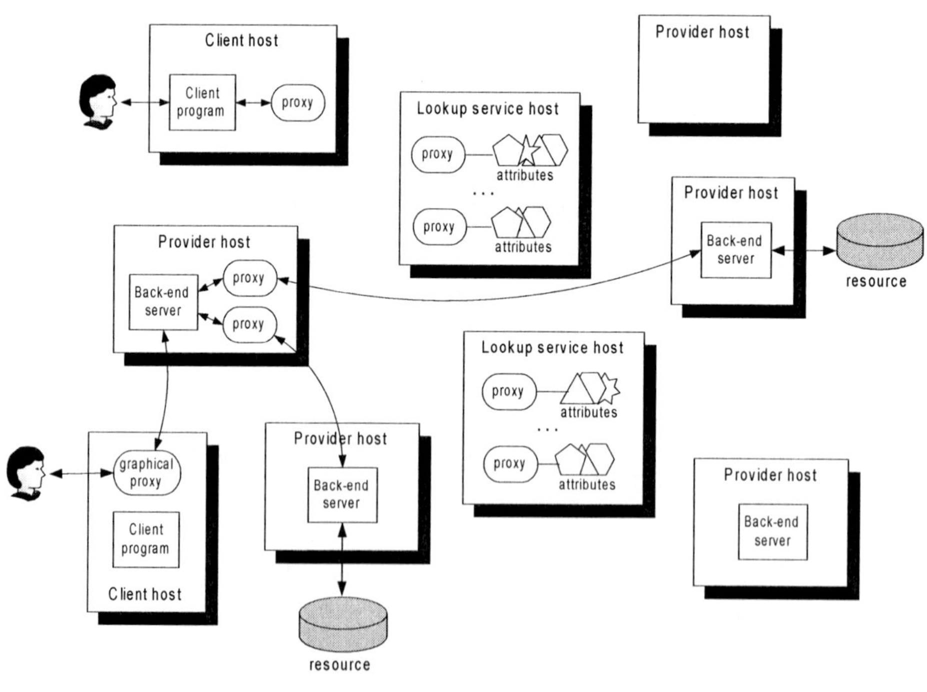

are not connected to each other. The two hosts manage two distinct registries that are not necessarily in accord, although can contain common records. Providers and clients can locate LUSs with the so-called "lookup discovery" protocols (Sun, 1999). Applying at least one of the three available lookup discovery protocols, providers and clients can obtain the proxies that belong to the LUSs of the djinn. (As it was mentioned earlier, a lookup service is an ordinary Jini service, so to use it first its proxy must be obtained.) Using the discovered LUS proxies, providers can publish their own proxies and suitable descriptor objects at these information system hosts. Clients can use LUS proxies to search in the database and to download service proxies. This searching procedure can be based on two different constraints:

1. Attributes the needed proxy (or proxies) has to be associated with.
2. Programming interface(s) the needed proxy (or proxies) has to implement.

While the first approach in itself can be enough to find proxies that implement user interfaces besides the programming one, the latter approach has to be applied if the user (or a high-level provider) wants to use the service through a client program. (Only the latter case can assure that the proxy will support the methods the client program is going to call on it.)

Jini, as a service-oriented middleware technology, meets the requirements of the Grid. In Jini networks resources can appear as services and lookup hosts can help clients and providers find each other. Everything is given for the on-demand, automated integration of resources. This has been graphically presented in Figure 10, since the user in the bottom of the figure can utilize multiple shared resources in an aggregated way. The two stateful resources (one on the right and one on the bottom) are wrapped into two distinct services. Resource wrapping solves the sharing problem, but not the resource aggregation. This aggregator functionality is provided by another service situated on the left. The core of this service is a server program that knows how to utilize the two services through their proxies. This "aggregator service" represents a new, virtual resource to the user. Since the proxy of this high-level service contains a graphical user interface, the user can exploit the capacities of two individual remote resources without needing any special client.

As discussed at the beginning of this section, Jini has been developed as a technology to enable the automated management of resources. Consequently,

Jini does not need any extensions to deal with stateful or dynamic providers. While Web services needed the OGSA tools and the OGSI/WSRF specifications to standardize the implementation of stateful services, Jini can cope with the issue alone. Jini contains every necessary tool to help developers handle the different aspects of dynamic and stateful behaviors. Without giving a detailed overview of the tools Jini provides to handle lifetime management, client notification, and all other related problems, we can declare that they are quite similar to the OGSA solutions, but they are implemented in Java instead of XML. Further information on Jini technology is given by Kumaran and Kumaran (2001).

Smart Organizations and the Grid

Researchers and developers have a lot to do before the Grid can become a reality. They first have to test the introduced concepts and principles in practice. Although there are both Web services and Jini networks that have been operating for a few years, they are not big enough to indicate how these systems would work on a global scale. We do not know what the most effective information system implementation is in this case, how many requests must be served by the middleware within a given period of time, and how much traffic will be generated between different departments, institutes, countries, or continents. Although these questions concern only to the bottom layer of the Grid, to the grid middleware, we are still unable to answer them.

The situation is even worse if we move to the next level, at which services are built onto the top of the middleware. What kind of services should be developed? What kind of resources should be represented as low-level services and what functions should be provided at higher-levels? What would be the most suitable interfaces for the different types of services? What patterns should high-level "superproviders" use to integrate low-level services into their own material and information processes? We have only hazy answers to these questions.

We are sure that this part of the research cannot omit active participation from the potential users. The users have to specify what kind of functionalities they would like to see in the Grid, what kind of functionalities should be wrapped into grid services. Once the users' preferences are known, researchers and developers can estimate what kind of logical and physical resources should be applied, and in which way, to provide the required functionalities. While the development of low-level services will be almost self-evident, high-level services will be quite involved to be created.

Since sooner or later the Grid will work on a financial basis, the real question is: What functionalities will users be willing to pay for? It is still a usual notion today that the Grid will be the source of "unlimited" computational power for everyone. Although it is true that computational power will be cheaper in the Grid era than it is today, the idea is basically false. A Grid, in which providers cannot get any money for the services they offer, cannot be stable and reliable. Although it is quite probable that basic grid services—for example the "scientific calculator" service used as an example in the previous section—will be offered for free, highly developed services that integrate dozens, hundreds, or thousands of devices and software components into a single virtual resource will be too expensive for an average user. Getting access to these high-level grid services will be the privilege of a few favored persons, just like today with supercomputers.

The Grid introduces a concept right on the top of network protocols that enables the representation of physical tools and abstract entities in a standardized way, as software services. In the service-oriented world an "intangible" networked entity, such as a knowledge base, can be described just like a "tangible" device: with a software interface. Nevertheless, high-level layers can be built only onto standardized lower ones. To get a highly intelligent system on the top of a standard service-oriented middleware, standardized low-level protocols are not enough. Services that implement custom interfaces require customized clients, so automated collaboration of devices cannot be carried out. Standardized service interfaces are needed that cannot be changed in the future. Only standardized service interfaces can guarantee higher-level interoperability, which is the key to defining virtual organizations.

The more companies support the interface standards, the bigger virtual organizations can be realized on top of the service infrastructure. Recall that a grid-based virtual organization specifies a set of persons and resources that have the possibility, and not just the ability, to collaborate with each other on demand. While grid protocols and patterns enable the sharing and aggregation of resources in countless different ways, virtual organizations restrict this with security (or even legal) constraints. While standards, defined at the network level, guarantee data interoperability, standards defined at the level of the Grid guarantee information interoperability (See Figure 11). Information interoperability guarantees that an XML document or a Java object has the

Figure 11. The connection between the knowledge pyramid (Filos & Banahan, 2000) and grid-based smart organizations

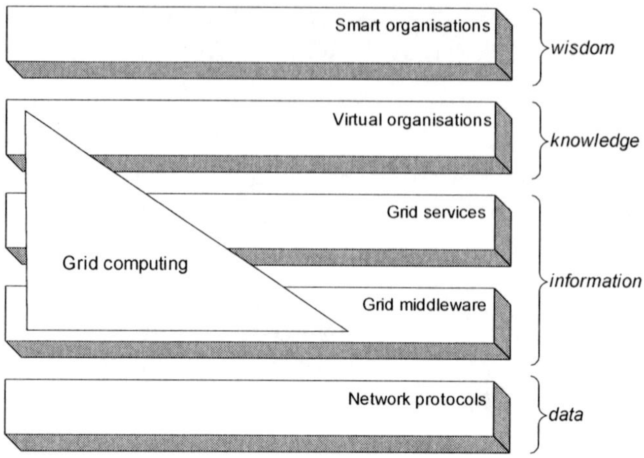

same meaning for both the sender and the receiver. Virtual organizations represent knowledge built onto the top of the service layer. This knowledge is realized as rules that determine what kind of partners the different services are allowed to interact with in order to fulfill their predefined tasks. While the grid middleware layer enables networked components to discover and connect to each other, realizing any pattern they would like to, virtual organizations force constraints on them. Although there is an opportunity to dynamically change these constraints, the "virtual organizations layer" in itself does not know when and how to do that.

This higher-level knowledge must be possessed by the layer built onto the top of virtual organizations, by the layer of smart organizations. This topmost knowledge is the "wisdom" necessary to control grid-based virtual organizations in a "smart" way.

"The term 'smart organisation,' is used for organisations that are knowledge-driven, internetworked, dynamically adaptive to new organisational forms and practices, learning as well as agile in their ability to create and exploit the opportunities offered by the new economy" (Filos & Banahan, 2000). While knowledge—stored inside the "virtual organizations" layer—can describe what the different components are allowed to do, the wisdom stored in the layer of "smart organizations" represents the past, the present, and the future of the entire underlying component set and the environment. On one hand, the wisdom a smart organization possesses is a synthesized version of its history. It contains information about the processes that have been taking place both inside and outside of the smart organization, and about the effects these processes had on the environment and on the different internal organizational layers. On another hand, wisdom includes the present status of the smart organization. A smart organization knows the current state of the components it contains and it is connected with. It knows even its environment, the physical world the whole service community operates in. The different low-level resources that participate in the smart organization can observe the environment with sensors attached to them, can filter and preprocess the information, and can send it up to the "brain" that controls the whole underlying structure. In addition, the wisdom of a smart organization determines its future. All the experiences collected in the past are all used by the smart organization to identify the opportunities and threats it will face with in the forthcoming periods. Using the synthesized information, the whole structure, controlled in a top-down fashion, can make itself a more profitable form, can achieve a better global optimum. Exploiting all the benefits of the flexible service-oriented infrastructure, the top layer can integrate new elements or parts of the network into the smart organization; it can disjoint unnecessary elements and redefine the connections among the already involved parts. Smart organizations, built onto the top of grid-based virtual organizations, can realize giant systems that are able to automatically change their structures and behaviors, extend or shrink themselves, and redefine their inner connections and the relations they have with their environment. The aim is to exploit the opportunities offered them by the physical world they are living in as efficiently as possibly.

Researchers and developers from various fields of sciences have to work together to turn smart organizations into a reality. Network technologists, security experts, artificial intelligence researchers, Web and multi-agent system developers, hardware engineers, and representatives of several other disciplines all have to be involved into this ambitious "project." Moreover, since smart organizations will directly or indirectly determine and control all the information and material processes around us, future research cannot miss participation from rather human fields of sciences such as economics, sociology, or jurisprudence. Experts from these significantly different fields of science have to add their own minor or major pieces into the big picture of smart organizations. Grid computing is contributing with a crucial part: with a service-oriented middleware that enables the on-demand sharing and aggregation of physical and logical resources.

For about a decade, Grid research has been focusing mostly on the development of the "grid middleware" layer from the architecture of Figure 11. While in the early years resource-oriented, today service-oriented middleware solutions are on the carpet. Besides finding the right middleware grid, researchers are trying to develop standardized services. Unfortunately, all these efforts have focused only on computational-related services so far. These services could realize the sharing and aggregation of processing power, storage space, and network bandwidths to serve the computational intensive jobs of e-scientists. Although the Global Grid Forum (GGF), the "standardization body" of the Grid community, has been working on the standardization of computational-related service interfaces for a few years, no widely accepted solutions have been created during this period. The most important reasons for this are probably the continuously developing middleware technologies, and the opposite interests of companies and institutes GGF is influenced by.

The third layer that is in the focus of the grid research community is the layer of the "virtual organizations." The results here are even less significant than in the case of the standardized services. The most widely used solution and tool for virtual organization management is still the Grid Security Infrastructure (GSI) (Foster et al., 1998) and its implantation contained by the Globus Toolkit (Foster & Kesselman, 1998). Since the GSI concept has been developed for resource-oriented Grids, quite probably it will not meet the requirements of flexible, service-oriented infrastructures.

We prognosticate that the triangle of grid research (see Figure 11) will shift toward the higher layers of the architecture in the forthcoming years. Even if it will not reach smart organizations, we are sure that significant research capacities will turn from middleware technologies to standardized services and service-oriented virtual organizations. The reason is that the industry already realized that the service-oriented architecture in itself provides several market benefits for them. With service-oriented IT infrastructures they can reduce installation and maintenance costs and can improve market competitiveness at the same time. Consequently, in the last few years service-oriented middleware research has become influenced by the industry rather than the grid community.

The triangle will not just move, but it will change shape as well. Its upper part will flare out, so at in the end it will become a convex trapezoid, or maybe a rectangle, or even a concave trapezoid. This transformation will express the tendency that grid researchers will develop services and virtual organizations for "ordinary" needs and not only for high performance and high-throughput computing purposes. While services of the latter type can serve e-scientist, services from the formal class can serve the "average user," who does not want to simulate earth quakes or study gene sequences. The "average user" would like to perform complex things, such as hotel reservation or video conferencing, in an easy and convenient way. Since high-level services, built onto a standard global middleware, would be able to cope with these issues too, we believe that grid research will open towards "the public" in the next future.

Summary and Conclusion

The first two sections of the chapter introduced the basic goal of grid computing. The aim is to create an infrastructure that physical and logical resources can apply to provide or discover functionalities on demand.

The second section presented the latest, service-oriented grid approach. To represent networked resources in a standardized way, all the hardware and software tools have to be wrapped into software services. Through the service-oriented middleware the participating resources can connect together to form high-level "superdevices." These new entities can join the Grid again, and the process can start afresh. As a result, the humans—the users of the top-level services—will be provided with virtual tools that represent resource capacities and software intelligence they have never seen before. To enable service-oriented collaboration, a suitable information system is required. The key question today is what technology to use to implement the information system, the services, and their clients. The technology has to be flexible enough to enable participation in any kind of hardware or software entity, but it should define behavioral and communicational patterns to guarantee service-level interoperability.

The third part of the chapter presented two service-oriented technologies. First, the Web services-based OGSI/WSRF technology has been examined, then the object-oriented Jini has been discussed. While Web services-based grids have the advantage of language independence, Jini grids can exploit the benefits of the stable Java platform that already proved its appropriateness in networked computing. The first option flatters with the promise to connect any kind of tool into the Grid, but it cannot provide the fancy features that the second does. If the Web services technology will be chosen, the Grid will probably get into our everyday life gradually, from one small step to another. Because of the strong relationship between today's Web technologies and the Web services platform, the average Internet user will not realize the coming of the Grid era. He or she will still contemplate and use the virtual world of the Internet through his or her well-known browser, without knowing that the resources that form this world cannot be regarded as individual computers any longer, they automatically cooperate with each other to serve the high-level demands. The situation does not get much better if Jini is chosen, either. Despite of the extra features it is able to provide, Jini networks will probably be accessed through Web portals, too. The Web already became such an integral part of our life that it cannot be changed to something else simply because a new technology provides a slightly better solution.

In the future, the two technologies will probably live parallel to each other. Some local area grids will apply Jini, while others will apply Web services. Although we can be sure that bridging solutions will be developed to make the collaboration of different local grids possible (it happens every time systems reach their boundaries), these bridging techniques will not be long-lived. Current trends in distributed computing unambiguously indicate that in the long run only one technology can be the winner. This technology will be "The Grid technology" and will be used to integrate together computers, electronic devices, vehicles, engines, and software components in a standard and generalized way. This flexible infrastructure can then be the foundation for smart organizations, for highly intelligent systems that can understand and influence our environment in a highly intelligent way.

References

Czajkowski, K., et al. (2004). *The WS-resource framework.* Retrieved October 10, 2005, from http://www.globus.org/wsrf/specs/ws-wsrf.pdf

Edwards, W. K., & Rodden, T. (2001). *Jini example by example.* Prentice Hall.

Filos, E., & Banahan, E. (2000). Will the organisation disappear? The challenges of the new economy and future perspectives. In L. M. Camarinha-Matos, H. Afsarmanesh, & R. J. Rabelo (Eds.), *E-business & virtual enterprises* (pp. 3-20). Dordrecht: Kluwer.

Foster, I., et al. (1998). A security architecture for computational grids. In *ACM conference on computers and security* (pp. 83-91). ACM Press.

Foster, I., & Kesselman, C. (1998). The Globus project: A status report. In *Proceedings of the Heterogeneous Computing Workshop* (pp. 4-18). IEEE Computer Society Press.

Foster, I., & Kesselman, C. (Eds.). (1999). *The grid: Blueprint for a new computing infrastructure*. Morgan Kaufmann.

Foster, I., & Kesselman, C. (Eds.). (2003). *The grid 2: Blueprint for a new computing infrastructure*. Morgan Kaufmann.

Foster, I., Kesselman, C., & Tuecke, S. (2001). The anatomy of the grid: Enabling scalable virtual organizations. In *Proceedings of the 7th Euro-Par Conference*.

Foster, I., et al. (2003). The physiology of the grid. In F. Berman, G. C. Fox, & A. J. G. Hey (Eds.), *Grid computing: Making the global infrastructure a reality* (pp. 217-250). John Wiley & Sons.

Fox, G., et al. (2003). Peer to peer grids. In F. Berman, G. C. Fox, & A. J. G. Hey (Eds.),*Grid computing: Making the global infrastructure a reality* (pp. 471-490). John Wiley & Sons.

Glass, G. (2002). *Web services—Building blocks for distributed systems*. NJ: Prentice Hall.

IBM. (2004). *WSRF announcement*. Retrieved October 10, 2005, http://www.marketwire.com/mw/release_html_b1?release_id=61977

Juhasz, Z., & Sipos, G. (2004). A Jini technology-based service grid. In ERCIM News Special: Grids: The Next Generation. No. 59, 29-30.

Kumaran, S. I., & Kumaran, I. (2001). *JINI technology: An overview*. Pearson Education.

Litzkow, M., Livny, M., & Mutka, M. (1988). Condor—a hunter of idle workstations. *Proceedings of the 8th International Conference on Distributed Computing Systems* (pp. 104-111).

Sun Microsystems. (1999). *Jini technology core platform specification*. Retrieved October 10, 2005, http://www.sun.com/software/jini/

Tuecke, S., Czajkowski, K., Foster, I., Frey, J., Graham, S., & Kesselman, C. (2002). *Grid service specification*. Global Grid Forum.

UDDI standard. Universal Description, Discovery and Integration. Available online at: http://uddi.org/

W3C, XML. *World Wide Web Consortium—Extensible Markup Language (XML).* Available online at http://www.w3.org/XML/

Waldo, J. (1999). The Jini architecture for network-centric computing. *Communications of the ACM, 42*(7), 76-82.

WS Activities. (2002). *World Wide Web Consortium.* Available online at http://www.w3c.org/2002/ws

Endnote

1 The Powerline Communication (PLC) technology is not considered here.

Chapter X

Communication Security Technologies in Smart Organizations

Raphael C. W. Phan,
Swinburne University of Technology (Sarawak Campus), Malaysia

Abstract

In this chapter, we discuss the security technologies that are important in guaranteeing the good quality of communication within smart organizations. We first briefly review the various forms of communication that can be used in the current information age, before outlining the possible threats that can be faced in each communication medium. We then describe the relevant security technologies that help to protect communication media from common threats, as well as the security tools available in the market that implement these technologies. The topics discussed in this chapter would serve to educate the smart organizations towards securing their various means of communication, which is vital for a business establishment to exist and coexist with peers and partners.

Introduction

Smart organizations are knowledge-driven, internetworked, dynamically adaptive to new organizational forms, agile in ability to create and exploit opportunities offered by the new economy (Filos & Banahan, 2000). Being internetworked, therefore, some form of communication has to exist between two or more parties. This communication has to be effective and dependable, and furthermore the parties would have to know what is basically happening behind the scenes, and be ever ready to upgrade their knowledge with the latest in technology. Otherwise, this may result in communication breakdowns and hence prevent transactions from being accomplished or contracts from being sealed properly with peers and business partners. What this means is dependability of the communication process, and is the focus of this chapter.

Dependability means that our system can be trusted to perform the service for which it has been designed, and can be decomposed into specific aspects as follows. *Reliability* characterizes the ability of a system to perform its service correctly when asked to do so. *Availability* means that the system is available to perform this service when it is asked to do so. *Safety* is a characteristic that quantifies the ability to avoid catastrophic failures that might involve risk to human life or excessive costs. Finally, *security* is the ability of a system to provide the following services (Stallings, 1999; Menezes, van Oorschot, & Vanstone, 1996) to communicating parties:

- **Confidentiality:** Ensures that the communicated information is accessible only by authorized parties.
- **Authentication:** Ensures that the origin of the message is correctly identified.
- **Integrity:** Ensures that only authorized parties can modify the communicated information, or enables parties to detect any unauthorized modifications to the information.
- **Non-Repudiation:** Ensures that neither party can deny having made any previous communications.

This chapter presents a discussion of security technologies available today to ensure the dependability of the communication process, which is vital within smart organizations since its parties are internetworked with each other, and therefore prone to network attacks and exploits by malicious crackers. One of the most important ways that smart organizations use to communicate is via the Internet. Performing transactions online via the Internet is an effective means (VeriSign, 2002) by which organizations can advertise and perform transactions with customers and other parties. However, online transactions will only be popular if the public trusts in their security (Amazon, 2003; Bolivia, 2003; Harris, 1998; Rawal, 2003; Tedeschi, 2000). Therefore, for an organization to be able to compete and advance, it needs knowledge, and hence careful management, of the various security technologies (Anderson, 2001; Garfinkel & Spafford, 1997) that help protect and safeguard public trust in its online transactions. The interested reader is also referred to the chapter on "New Challenges in Smart Organizations: Demands of Mobility" that also appears in this book for a discussion of other relevant future trends.

At the end of this chapter, we hope that the reader would have obtained a general perspective of communications security technologies that can be used in smart organizations. In particular, the objectives of this chapter include:

- Understanding of the various types of communication techniques
- Understanding of the possible threats faced by the communication process
- Familiarity with communication security technologies such as *encryption*, *digital signatures,* and *message authentication codes (MACs)*
- Familiarity with common software and hardware tools used to provide security technologies

Background

Security is an important criteria these days, especially with in current information age in which information is available and accessible everywhere, in any form and with any means. The largest depository of information is the Internet, where infinite information is speedily available at one's fingertips.

With this vastness and freedom of information also comes the threats of abuse and misuse by malicious parties whose intent could be to deceive, steal, impersonate, cheat, or merely intrude into others' privacy.

Smart organizations are equally affected by this, since the communication process within such establishments requires speed, ease of access, and wide coverage, and further they have to readily adapt to using the latest communication techniques such as the Internet, high-speed networks, and wireless communications. Nevertheless, one has to do so with caution since new technologies are at times not fully tested, and may further have bugs exploitable by attackers to gain entry into the organization's network. Also, smart organization personnel should each be fully aware of how to use such technologies, the possible weaknesses inherent in them, plus how to prevent attacks of any kind.

Protecting the security of the communication process is indeed very important. Based on the recent 2004 Computer Crime and Security Survey (CSI, 2004) by the U.S. Computer Security Institute (CSI) and FBI, it was shown that the Internet connection is increasingly becoming the most frequent point of attack, mostly due to viruses, abuse of net access, unauthorized access, system penetration, and denial of service. Other popular attacks include sabotage, financial fraud, and telecom fraud, which also target the communication process.

Ways of Communication in Smart Organizations

In this section, we will look at the ways in which two or more parties can communicate within smart organizations. The main characteristic of communication these days is that it is mainly done via the internet or wireless networks. In other words, communication is online.

World Wide Web (WWW)

One of the earliest ways of communicating online at high speed was via the World Wide Web (WWW), which incorporates all Internet services and allows various parties to access documents, images, music, videos. These are all hosted on computers that run special software called server software and are online all the time. One simply uses a Web browser software such as Netscape's Navigator or Microsoft's Internet Explorer in order to view these online information. Websites in the WWW start with the header "http://www"—for example, http://www.google.com.

File Transfer Protocol (FTP)

Another popular way of communication that has been popular since some time ago is via the file transfer protocol (FTP). This is a specially dedicated specification of how one could transfer files from one computer to another. This is similar to sharing folders or directories in your computer with everyone in the world, and they can log in and access them as if they were physically sitting in their own computers. FTP sites can also be accessed via web browsers or special FTP client software. FTP addresses start with the header "ftp://"—for example, ftp://ftp.example.com.

Electronic Mail (E-Mail)

By far the most popular and commonly-used communication technique is the electronic mail (e-mail). These are messages sent through the Internet from one party to another. Not only that, but other files such as images, music, and video can be included as attachments to these e-mails. Every party communicating via e-mail has an e-mail address, with the format username@emailserver.domain—for example rphan@swinburne.edu.my.

Telnet

Telnet is a terminal emulation program for the Internet. This means that when you run a telnet program on your computer, you can connect your computer to another telnet server on the Internet so you can interact with it by typing in commands, as if you were physically sitting at that server machine and typing at its keyboard. This allows you to remotely control computers and servers.

Chat

Internet chat is popular especially among the young and young at heart. You could chat either using special chat software such as ICQ or Internet relay chat (IRC), or on specific chat Web sites. A chat window is open throughout the chat period and you can type your message and the person at the other end would almost immediately see it. Advanced chat features also include chats among more than two parties (called meetings or conferences), voice chats, and streaming video of the chatting parties.

Short Message Service (SMS) and Multimedia Message Service (MMS)

Meanwhile, the mobile phone revolution started several years ago and never looked back. These days, almost everyone has a mobile phone; for instance, the 2004 mobile penetration rate (Netsize, 2005) in European countries including the UK, Italy, and Sweden has crossed 100%. Although mobile phones were initially used mostly for voice conversations, another revolution ensued, namely the Short Message Service (SMS). This allows parties to send short text messages (up to 160 characters) to one another by using their mobile phones. Following this is an advanced version known as the Multimedia Message Service (MMS), which also allows short videos—including sound—to be sent as messages to other mobile phones. This is just starting to become popular, and is also envisioned to revolutionize the way people communicate in future.

Threat Models for the Communication Process

Threats faced by communicating parties are generally grouped into threats from two types of attackers: namely, passive attackers and active attackers (Stallings, 1999).

Passive attackers are those who eavesdrop on or monitor the communications channel, but do not affect or interfere with the communication in any way. Therefore, such attackers are very hard to detect since there is no straightforward way of knowing when communication is being monitored.

Considering the previously discussed ways of communication, passive attacks on the WWW, FTP, and e-mail services, for example, could involve simply monitoring which addresses (WWW, FTP or e-mail) that a certain party is accessing or communicating with. This is possible because these communication services by default simply involve the communication of messages that are in the clear, meaning in readable form. So an attacker could obtain information related to the personal life of a party—for example, which types of Web sites he has visited,what type of files he has accessed, what e-mail addresses he sends to, and so forth. This of course intrudes on the privacy of communicating parties. Individual communicating parties are often not aware of such threats, and do not bother to customize default security and privacy settings within the communication software they use.

Mobile communications via mobile phone conversations, SMS, and MMS are also potentially susceptible to passive attacks. Messages are communicated via microwave signals, which can be easily intercepted by specially built receivers that tune in to the same frequency. Further, since these signals are free to propagate through the air, they are out in the open and so it is almost impossible for one to detect if they are being read by passive attackers. This is in contrast to a telephone line, where a passive attacker would have to make a physical tap somewhere along the line in order to do any eavesdropping. Therefore, there is a need in mobile communications to cover up the meaning of the message. This is usually achieved via the technology called encryption, which will be discussed in the next major section. Doing so ensures that even if a message were intercepted, a passive attacker would not be able to understand what it is.

Active attackers, on the other hand, are those who directly interfere with the communication, either by interrupting, modifying, or fabricating messages. *Interruptions* of messages are direct attacks on the availability of the communication service, while *modifications* are attacks on the integrity of the service. Finally, *fabrications* are attacks on the authenticity of the communication service. All these are serious attacks and should be guarded against.

An active attacker would be able to mount more devastating attacks on the above communication services. For example, he or she could modify the e-mail messages being sent from one party to another and hence cause reputation-damaging consequences. Since e-mails are by default sent in the clear, spoofing such as this can be done easily. Similarly, he or she could masquerade as a certain Web site and dupe WWW browsers into providing information that he is not supposed to have; for example, passwords and credit card information. This is known as "phishing," and is a huge threat because there is a serious lack of public awareness on this. The reason is partly also due to most parties being unfamiliar with the latest technologies, such as the WWW. Therefore, they do not know what is really happening and simply trust such Web sites for what they claim to be. For this reason, commercial banks are starting to give warnings to their customers not to trust any Web site claiming to be them (Personal Computer World, 2004) because Web sites are so easily set up. Another common type of active attack is the Denial of Service (DoS) attack on e-mail servers and WWW servers that intentionally makes limitless accesses to a communication service and heavily overloads the service provider until it disrupts the provision of service to authorized and legal customers. What aids this is the fact that unsuspecting computer users could be forwarding interesting e-mail attachments to friends and colleagues that contain Trojans—malicious software that appear innocent but are doing malicious activities in the background. Such Trojans can be used to collectively mount DoS attacks on Web sites.

Another example of active attack is hacking into a network via a computer within the network, such as through the telnet service provided by the computer to outsiders. Aiding the increase of such hacks is the fact that the learning curve for hackers is decreasing with the vast availability of easy-to-use hacking tools freely downloadable from the Internet (Yunos, 2004).

Finally, viruses and worms are active attacks and really devastate an organization's networks and operations. The damages done could cause losses in millions of dollars (Yunos, 2004) to both the government and the affected organizations themselves. Viruses and worms spread easily due to individual computer users' unawareness of how to safeguard their own computers against these threats. Often, these threats enter an organization's network via an innocent and ignorant employee's computer. The easy availability of virus authoring tools also means new variants can be speedily created by amateurs, and may escape detection by antivirus software.

Essential Security Technologies in Smart Organizations

The most fundamental (primitive) security technologies (Stallings, 1999) that allow one to ensure the security of communication are known as encryption, digital signatures, and message authentication codes (MACs). These fundamental security technologies provide the basic security services of confidentiality, authentication, integrity, and non-repudiation described in the preceding introduction section. We also include in this section a discussion of security protocols and standards that make use of these security primitives.

Encryption

Encryption is basically the process of transforming an original confidential message (plaintext) into unreadable form (ciphertext), and is done by using a key, often generated from a password that is specified. Only the persons who know the key can reverse the process (decrypt) to get back the original message. This clearly ensures that only those who are supposed to read the message can do so. Encryption provides the confidentiality service.

The method to do encryption is called an algorithm. There are generally two types of encryption algorithms, namely symmetric and asymmetric algorithms. Symmetric algorithms use the same key for doing encryption and decryption. This is shown in Figure 1.

By using a secret key, the original message (plaintext) is encrypted into ciphertext. Once in this form, someone who happens across it would not be able to discern its meaning. In order to get back the original message, the same secret key is used to decrypt the ciphertext (Refer again to Figure 1).

One problem with symmetric algorithms though, is how to share the secret key with others whom you trust. Suppose you want to send an e-mail to your wife. By using a symmetric algorithm, you encrypt your email with a secret key. Then

Figure 1. (a) Symmetic encryption with a secret key; (b) symmetric decryption with the same secret key

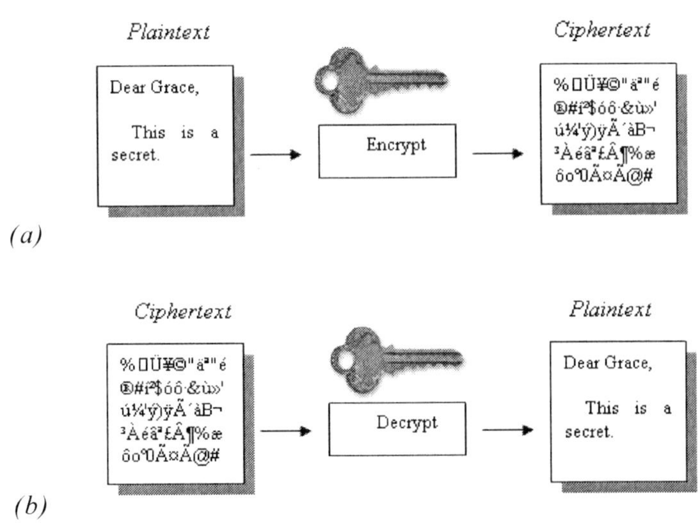

you send the encrypted email to your wife. The problem is how to let her know what is the value of the secret key that you used? You could call her up and whisper it over to her on the phone, but what if the phone line is tapped into? E-mailing the secret key to her is pointless because the fact that you do not trust the telephone line was why you wanted to encrypt your email in the first place.

Asymmetric (public-key) encryption algorithms overcome this problem by using two keys, called the public key and the private key. Each person has his own public key and private key. When you want to send an e mail to your wife, you use her public key to encrypt the e mail. Then when your wife wants to read the e-mail, she simply uses her private key to decrypt the e-mail. This is shown in Figure 2.

Anyone else who tries to decrypt the e-mail would obtain nonsense because only one private key, in this case your friend's private key, can correctly decrypt the e-mail. Your friend can tell everyone what his or her public key is so that anyone who wants to send e-mails to him or her can use th public key to encrypt the message. But knowing the public key does not help an unauthorized person to decrypt the e-mails because only his or her private key can do that.

Obviously, asymmetric encryption algorithms seem more desirable than symmetric encryption algorithms. However, they tend to be slower than symmetric

Figure 2. (a) asymmetric encryption with the public key of the recipient; (b) asymmetric decryption with the private key of the recipient

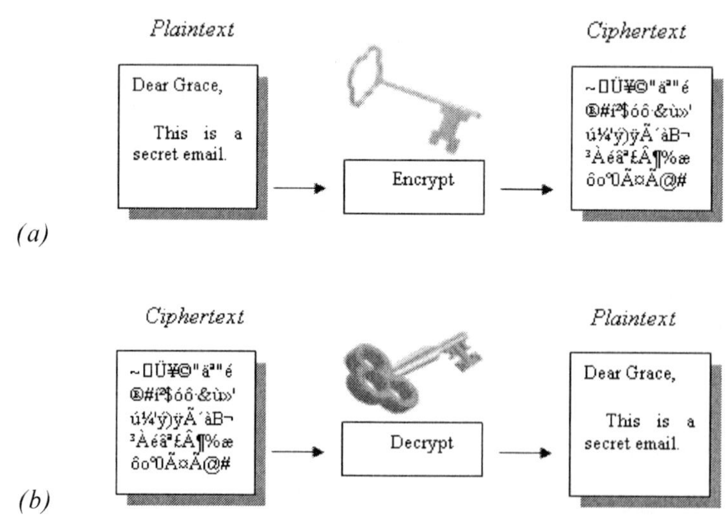

Figure 3. Encryption by the sender

(a) Symmetric encryption of the e-mail with a secret session-key, K

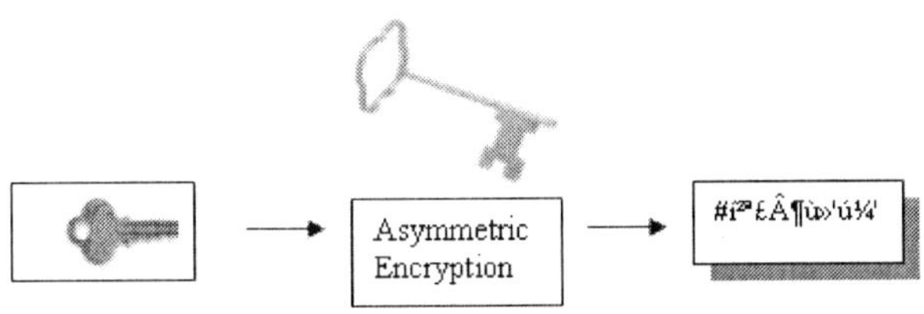

(b) Asymmetric encryption of the secret session-key, K, with the public key of the recipient

algorithms, especially when the message to be encrypted is large. Due to this fact, the best solution then is to combine the advantages of both methods, as is the current practice by most e-mail security standards.

Whenever an e-mail message is to be sent, the following steps are taken, as illustrated in Figure 3:

1. A random secret session-key, K, is generated.
2. The email message is encrypted by a symmetric encryption algorithm with the secret session-key K.
3. The session-key K is encrypted using the recipient's public key.
4. The encrypted session-key K and the encrypted e-mail message are sent to the recipient.

Both the encrypted e-mail message and the encrypted session-key K would be sent together over the Internet as one e-mail to the recipient. This is shown in Figure 4.

Figure 4. Sending both the encrypted e-mail and encrypted session-key over the Internet

When the e-mail is received by the recipient, the following is done, as in Figure 5:

1. The recipient uses her private key to decrypt the encrypted session-key, so that she can get back the session-key K that was used to encrypt the e-mail message.
2. She uses the session-key K to decrypt the e-mail message.

An e-mail message could be quite large, so it is encrypted by using symmetric encryption because symmetric encryptions are faster than asymmetric encryptions. Then, in order to solve the problem of having to inform the recipient of the value of the secret key used in the encryption, asymmetric encryption is used to encrypt the secret key. Since the secret key is small compared to the email message, it does not take up much time.

Digital Signatures

A *digital signature* is a way to ensure that an electronic document (your e-mail or MS Word document, for example) is authentic. Being authentic means two things:

Figure 5. Decryption by the recipient

(a) Asymmetric decryption of the session-key, K, with the private key of the recipient

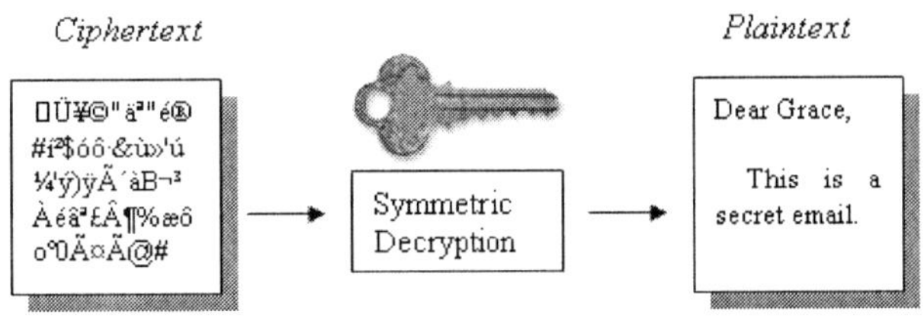

(b) Symmetric decryption of the e-mail with the secret session-key, K

1. You can verify the identity of the person who sent the document
2. You can verify that the contents of the document have not been altered by a third party in any way since it was written

This makes a digital signature analogous to its handwritten counterpart in the pen and paper world. Digital signatures therefore provide the authentication and non-repudiation services.

A popular method of making use of digital signatures is through a *public-key encryption* system. In such a system, each user has two keys, a *public key* and a *private key*. The public key is widely known to the public. However, the private key is secret and known only to the user.

Most of the time, since documents are quite large in size, they are first *hashed* to produce a compressed version of the document, called the *message digest*. The message digest is encrypted (signed) with *your private key*, and this will be your digital signature. The digital signature is then appended to the end of the document, as illustrated in Figure 6.

Figure 6. Digitally signing a document

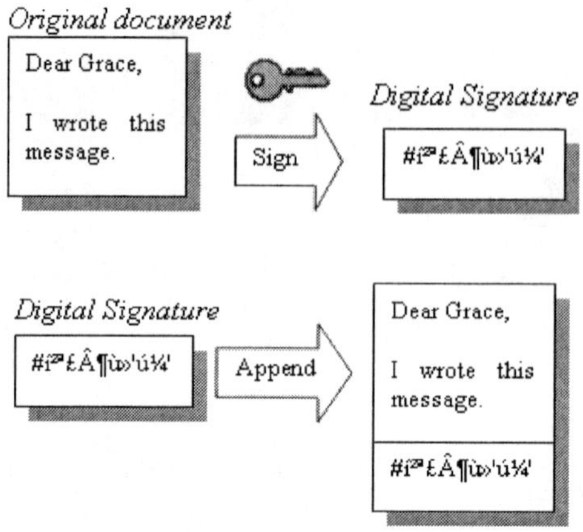

The document, along with the appended digital signature, is then sent to the recipient, Grace. To verify the document, Grace first hashes the document to get the message digest. She then uses *your public key* to decrypt the appended digital signature. The decrypted digital signature should be the same as the message digest that she obtained by hashing the document (See Figure 7).

The fact that the digital signature is correctly decrypted into the message digest verifies that the document was written by you, since it was signed by using your private key, which only you will have. If the message digest obtained by hashing the document is the same as the decrypted value of the digital signature, it proves that the document has not been altered since it was sent, otherwise a different message digest would be obtained if you hash it.

Notice how elegant a public-key encryption system is. You can have encryption or digital signatures depending on the way you use the public and private keys of the sender and recipient. It can be summarized as follows:

- For encryption, the sender encrypts with the recipient's public key. The recipient decrypts it back with her private key.
- For digital signatures, the sender encrypts with his own private key. Anyone who wishes to verify the digital signature uses the sender's public key to decrypt it.

Figure 7. Verifying the authenticity of a signed document

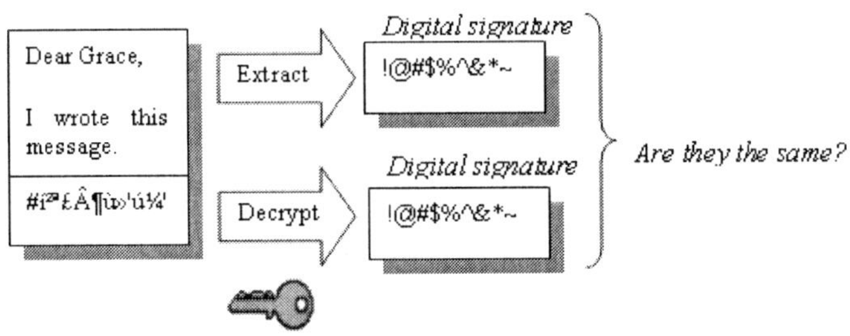

Encryption of a document with the intended recipient's public key ensures that anyone can encrypt and send to the recipient since his public key is publicly known, and is meant to be so. Decryption of that document can only be done with the recipient's private key, and since this key is only known to the recipient, no one can decrypt back the document except the intended recipient himself.

As for digital signatures, a document to be signed is encrypted with the sender's private key. This ensures that only he can sign since his private key is known to him alone. On the other hand, anyone can verify whether the digital signature is authentic by using the sender's public key, which is again widely known.

Message Authentication Codes (MACs)

Message authentication codes (MACs) are used to ensure the integrity of messages communicated between two parties in order to prevent active attackers from modifying or tampering with them in any way. A MAC works similarly as the secret-key encryption since it also depends on a secret key. The only difference is that it generates a small encrypted output, called a MAC, in contrast to normal encryption where the output is the same size as the input.

Let's assume that two parties, Alice and Bob, share a common secret key. Then when Alice wants to send a message to Bob, she inputs the message into a MAC generator and also puts in the secret key. The result is a MAC with a size much smaller than the original message. Alice next appends this MAC to the message and sends it to Bob.

When Bob receives this, he uses the same secret key to regenerate the MAC from the received message and then compares this with the MAC he received from

Alice. If they are the same, this means that the message has not been modified while it was being sent over, since if an attacker had done so, the two MACs would be different. The attacker would not have been able to alter the MAC to correspond to the modified message since he does not know the secret key used to generate the MAC. Therefore, this guarantees the integrity of the message.

Security Protocols and Standards

A protocol is a formal sequence of steps taken by two or more communicating parties. A security protocol specifies what the parties do in order to communicate securely. Security protocols make use of the primitives, Encryption, digital signatures, and MACs, in order to achieve this goal.

The de facto standard protocol for securing online transactions (i.e., buying things from the Internet, sending confidential information, etc.) is the Secure Sockets Layer (SSL) by Netscape Inc (Anderson, 2001). It allows a Web browser client and a visited Web server to authenticate each other, and to further exchange a secret session key for secure communication of sensitive information between them. Browsers secured with SSL would have a header of "https" instead of the normal "http." The Internet Engineering Task Force (IETF) is in the process (Stallings, 1999) of formalizing a similar standard, the Transport Layer Security (TLS), which is based on the SSL protocol.

The standard for secure credit card transactions on the Internet is the Secure Electronic Transaction (SET) standard developed by a consortium including Microsoft, Netscape, VISA, and MasterCard (Anderson, 2001; Stallings, 1999), basically allowing customers and merchants to authenticate each other via digital signatures, and to communicate confidentially via encryption. SET nevertheless does not seem to fully penetrate the market (Anderson, 2001) due to high costs and impracticality, and lack of response from customers.

A public key infrastructure (PKI) is basically a term given to systems that use public-key encryption and digital signatures, where each user (customer, merchant, etc.) has a pair of public and private keys, and where each has a digital certificate certified by a trusted Certification Authority (CA), containing the public key and personal details while the private key is kept secret. A party, Alice, wishing to communicate with another party, Bob, would access Alice's digital certificate for her public key, which can then be used to encrypt confidential information for Alice, or to verify digital signatures signed by Alice. In principle, the above SSL and SET protocols can be viewed as PKIs.

Tools for Security Technologies in Smart Organizations

The tools by which smart organizations could provide security technologies such as those described in the previous section could generally be classified into software-based and hardware-based approaches.

Software-Based Tools

The most common software-based tool that provides security services are security software developed by security-based software companies such as anti-viruses, or those that allow computer users to perform encryption, digital signatures, or MACs—for example, the Pretty Good Privacy (PGP) software (Stallings, 1999). Anti-viruses are commonly installed at individual computers and used to scan files commonly prone to viruses. Most anti-viruses these days are also able to be triggered into execution and remain in the background, monitoring the file activity and warning the user of suspicious actions by those files. Meanwhile, security software that provides encryption, digital signatures, and MACs are very useful in that they allow the computer user to protect the confidentiality of his files, authenticate other users, or check the integrity of received files for possible modifications. This software can also be used in relation to e-mail client software and hence help to provide these security services in e-mails, as well.

Another type of anti-virus software is the virus filter (Stallings, 1999), which sits at the server computer and scans files passing through the server for possible virus infections before they enter into the network. This helps to reduce the spread of viruses considerably, since virus filters are maintained by system and network administrators who have the technical know-how on viruses, in contrast to normal anti-virus software sitting on each computer, typically managed by users who sometimes even disable the real-time virus scanning feature in exchange for higher efficiency.

Other such security software tools include firewalls (Stallings, 1999). A firewall is a program running at a network gateway server that protects the files and other resources of the network from other outside networks. Such tools are especially important against active attackers such as hackers or crackers to prevent them from gaining access to computers within the network.

When one connects to a remote computer, the common Telnet session is known to be insecure as messages are transmitted in the clear. To combat (Anderson, 2001) network attacks and hacks, a secure alternative is the secure shell (SSH), which encrypts all messages and so whatever info such as login passwords that are transmitted will be save from eavesdroppers. Internet users also have the option of using encryption and authentication at the Internet Protocol (IP) layer, provided as IPSec in the IPv6 (IP version 6). Yet another alternative is using virtual private networks (VPNs), where several companies communicating with each other would have their firewalls arranged with each other to encrypt all traffic between them. Having said this, though both IPSec and VPNs prevent external network attacks, they do not tackle the issue of insider attacks, and so in general they should be used along with a complementing tool (such as multi layers of conditional access, security logs, frequent updates of keys) to also prevent insider attacks.

Hardware-Based Tools

Hardware-based security tools are gaining popularity these days, including password systems, magnetic stripe cards, smart cards, and biometric systems.

Password systems (Anderson, 2001) are the oldest hardware-based systems and these include those such as safes in banks and locks on briefcases. More advanced versions of these are those such as the automatic teller machines (ATMs) that employ both passwords and magnetic stripe cards to authenticate the legal party. Old versions of credit cards also use magnetic stripes for reading and verification. Nevertheless, the recent widespread cases of credit card cloning (Yunos, 2004) has raised concern on the security of magnetic-striped based cards.

Smart cards (Anderson, 2001) have been increasingly used in recent years in place of magnetic stripe cards since they are tamper-resistant and also contain a built-in microprocessor that processes and executes programs from within the smart card. The tamper-resistance of smart cards also appear to complicate active attackers from cloning them and using them for impersonations. The physical tamper-resistance in smart cards are usually implemented by using tamper-detection wires that immediately cause the memory within smart cards to be erased the moment illegal tampering is detected. Nevertheless, smart cards can still be hacked or cloned, though with a bit more effort compared to magnetic stripe cards, as clearly detailed in Anderson (2001).

Biometric systems (Anderson, 2001) are also widely used currently, and these make use of a person's fingerprints, eye pattern, face patterns, voices, hand geometry, and even typing behavior. These eliminate the need to carry a foreign object such as a smart card, since a person needs only to use his or her own body for authenticating himself to the system. Some important security issues (Anderson, 2001) regarding the proper use of biometrics include:

- **False positives and false negatives:** The former means a fake biometrics being falsely taken to be true, while the latter means a true biometrics being taken as a fake one. These are mostly due to careless biometrics laboratory procedure.
- **Freshness:** It is hard to tell the age of a biometrics print, and therefore there are issues of how fresh it is, or if it could have been planted there by someone else, since most biometrics such as voice, eye patterns, and fingerprints can be recorded and replayed at a later time.

In summary, it is preferred to use various tools such as smart cards and biometric systems in combination for extra security, and such that compromising one would still not compromise the entire system.

Future Trends

The introduction of the computer caused a digital revolution where vast amounts of information are expressed, communicated, stored, and processed in digital form. This revolution was also propelled considerably by the popularity of the Internet. This was followed by the mobile phone revolution, leading to the popularity of instant messaging via mobile communication networks. The trend is emerging where multimedia-based services via mobile devices would be at the forefront of the communication process. Also looking very prospective is the trend toward interactivity in communication; for example, viewers being able to interact with cable TV services via their televisions, mobile service subscribers downloading packages onto their devices, customizing their devices, and even more flexibility and customization by the car owner in the latest designs of high-tech cars. With each new communication trend comes new potential threats against the communication process, hence this area of communications security is an ever-changing field that improves over time and would never remain stagnant.

Conclusion

In this chapter, we first discussed the communication process and various forms in which it can take place. We then proceeded to the possible threats that might be faced by the communication techniques and then described the basic security technologies that could be used to guard against the attacks. This was followed by a discussion of the different software and hardware tools used to provide the previously-mentioned security technologies. The concepts discussed in this chapter are essential since an appreciation of these concepts would help the reader to better understand how the security technologies can be provided in the communication process, and also help personnel (both the leaders and subordinates) in smart organizations to manage and handle such security technologies better.

References

Amazon.com. (2003a). *Help/ordering from Amazon merchants/credit card security*. Retrieved March 4, 2003, from http://www.amazon.com/exec/obidos/tg/browse/-/468494/102-8032379-5145751

Amazon.com. (2003b). *Help privacy & security*. Retrieved March 4, 2003, from http://www.amazon.com/exec/obidos/tg/browse/-/551434/ref=br_lr_/102-8032379-5145751

Anderson, R. (2001). *Security engineering: A guide to building dependable distributed systems*. New York: John Wiley & Sons.

Bolivia Web. (2003). *Bolivia Web—the book club: Security & Amazon.com*. Retrieved March 4, 2003, from http://www.boliviaweb.com/books/security.htm

CSI. (2004). *CSI/FBI computer crime and security survey*. USA: Computer Security Institute (CSI). Retrieved March 28, 2005, http://i.cmpnet.com/gocsi/db_area/pdfs/fbi/FBI2004.pdf

Filos, E., & Banahan, E. (1997). Will the organization disappear? The challenges of the new economy and future perspectives. In L. M. Camarinha-Matos, H. Afsarmanesh, & R. J. Rabelo (Eds.), *E-business & virtual enterprises* (pp. 3-20). Dordrecht: Kluwer.

Garfinkel, S., & Spafford, G. (1997). *Web security and commerce*. O'Reilly & Associates.

Harris, L. (1998). *E-commerce and privacy survey*. Retrieved September 15, 2002, from http://www.msnbc.com/msn/427057.asp

Menezes, A., van Oorschot, P., & Vanstone, S. (1996). *Handbook of applied cryptography*. CRC Press.

Netsize. (2005). *Global mobile operators suffering from declining voice revenues look at mobile data services to come to the rescue*. Retrieved March 28, 2005, from http://www.netsize.com/?id=6&sid=4&rid=23

Personal Computer World. (2004). *OFT trawls uncovers 176 rogue Web sites*. Retrieved March 28, 2005, from http://www.pcw.co.uk/news/1154005

Rawal, D. (2003). *Response to Query on How Amazon.com Provides Online Security Services*. Personal communication via e-mail: form-info@amazon.com.

Stallings, W. (1999). *Cryptography and network security*. NJ: Prentice-Hall, Inc.

Tedeschi, R. (2000). *Sellers hire auditors to verify privacy policies and increase trust*. Retrieved September 15, 2002, from http://www.nytimes.com/2000/09/18/technology/18ECOMMERCE.html

VeriSign. (2002a). *Building an e-commerce trust infrastructure: SSL server certificates and online payment services*. Retrieved September,15, 2002, from http://www.verisign.com

VeriSign. (2002b). *Increase your competitive edge with a secure Web site*. Retrieved September 15, 2002, from http://www.verisign.com

Yunos, Z. (2004, November 5). *Protection on Malaysian cyberspace: An agenda for action*. Swinburne University Research Seminar Series, Malaysia.

About the Authors

István Mezgár is a senior researcher and deputy head of CIM Research Laboratory at the Computer and Automation Research Institute, Hungarian Academy of Sciences. He is an associate professor at the Department of Production Informatics, Management, and Control, Budapest University of Technology and Economics, Hungary, (Diploma in ME, 1977, Technical University of Budapest (TU); CSc in Tech.Sci., 1995, Hungarian Academy of Sciences). He is an editorial board member of numerous international journals, member of IFIP WG 5.3, and has more than 120 scientific publications at international forums. He served as visiting scientist and professor several times at different universities and research institutes of Italy, Japan, New Zealand, and South Korea. His current interests focus on networked organizations, wireless communication networks and their security, trust building, and smart card technology.

* * *

Peter Bertok is an academic staff member at the School of Computer Science, Royal Melbourne Institute of Technology, Australia. He received a PhD from the University of Tokyo, Japan, and a master's degree from the Technical University of Budapest, Hungary. He has authored more than 80 refereed publications. His main research areas are networked and distributed computing, Web services ,and computer security.

Shirley Chan started her career as a lawyer specializing in information technology law and moved into information and knowledge management. She was a business analysis consultant (information management) with the New Zealand government, a co-founder of the New Zealand Knowledge Management Network and an associate lecturer in information systems at the University of Melbourne in Australia. She is now a senior research associate at The University of Hong Kong Business School and a management consultant to a company within the Swire Group, which is Cathay Pacific's major shareholder. Shirley graduated from Hong Kong, Harvard, and Victoria Universities. She enjoys gym exercise, cooking, and spending time with her family.

Erastos Filos was born in Athens, Greece. After obtaining his MSc in physics from the University of Hamburg ,he worked as a research scientist at the High-Energy Physics Lab (DESY) in Hamburg, Germany. After that, he obtained his PhD from Constance University, Germany ,and started his business career, running a manufacturing project at Perkin Elmer Corp. and leading the electronics design team at Bosch Telecom. After 15 years of direct involvement in university and industry research and development, he joined the European Commission's Research Program in Brussels in 1993. His activities in research management were on "computer-integrated manufacturing and engineering," particularly in the domains "concurrent engineering" and "product modelling." Since 2003, he has been responsible for research strategy within the Department of Components and Systems, which is part of Information Society Technologies (IST) Research. Together with the department's director, Dr. Rosalie Zobel, he defined the research domain "Smart Organizations." They have also been successful in setting up an international cooperation pilot between IST and the National Science Foundation's Information Technology Research Program in the U.S.

Nathalíe Galeano is research assistant at the Center of Innovation in Design and Technology (CIDYT) of ITESM University, Monterrey, Mexico. He received a master's degree in manufacturing systems at the same institute (2002). He is a certified manufacturing technologist (CMfgT), Society of Manufacturing Engineers (SME), Dearborn, Michigan (1998). As research assistant, he has participated in consulting and training projects related to entrepreneurial management, enterprise modeling, benchmarking, concurrent engineering, development of virtual industry clusters, and collaborative networks organizations. He has work experience as a consultant in CEIS Company (Monterrey, Mexico), in supply chain management processes, logistics, productivity, and industrial engineering projects. He is also the coordinator of ECOLEAD Project.

Ricardo Mejía worked as research assistant at the Integrated Manufacturing System Research Center (CSIM) of ITESM University, Monterrey, Mexico (1999-2004). He received a master's degree in manufacturing systems at the same institute (2003). He is currently doing his PhD in the Laboratory LIPSI (Laboratoire en Ingénierie des Processus et des Services Industriels) in the engineering school ESTIA, France, together with the Wolverhampton University, UK. Gutiérrez has participated and coordinated research, consulting and training projects in the areas of concurrent engineering, information modelling for design and manufacturing, knowledge-based engineering, integrated product development, supply chain management, and virtual organizations.

Péter Kacsuk is head of the Laboratory of the Parallel and Distributed Systems in the Computer and Automation Research Institute of the Hungarian Academy of Sciences. He is a part-time, full professor at the Cavendish School of Computer Science, University of Westminster. He served as a visiting scientist or professor several times at various universities of Austria, UK, Germany, Spain, Australia, and Japan. He has been published two books, two lecture notes, and more than 160 scientific papers on parallel computer architectures, parallel software engineering, cluster and Grid computing tools. He has participated in and led several Hungarian and international grid projects since 1992. He is a co-editor-in-chief of the *Journal of Grid Computing*.

Arturo Molina is dean of the Engineering and Architectural Division of Monterrey Institute of Technology (ITESM), Campus Monterrey, Mexico. He earned a PhD in manufacturing engineering (Loughborough University of Technology, UK), a University Doctor degree in mechanical engineering (Technical University of Budapest, Hungary), and an MSc in computer science (ITESM, Mexico). Molina is a member of the National Researchers System of Mexico (SNI-Nivel I), a senior member and CEI of SME, IFAC Chair of WG

5.3, and a member of IFIP WG 5.3, IFIP WG5.12 and United Nations Task Force on ICT. Founder of IECOS company (www.iecos.com). His research interests include: information systems support for concurrent engineering, virtual enterprises, enterprise modeling and integration, information modeling for design and manufacturing, technology management, benchmarking, and technologies for global manufacturing.

Teresa Nájera received her master's degree in engineering sciences with a focus in manufacturing systems from ITESM, Monterrey, Nuevo León (2002), Mexico. She earned a BS in industrial and systems engineering with a focus in quality systems (1997) from ITESM, Torreón, Coahuila. She was a project engineer with CEMEX-SIPSA, Monterrey, Nuevo Leon (August 2002-September 2003). She was a project leader with PROLEC-GE, Apodaca, Nuevo León (March 2000-May 2001). She has worked on automatization of planning, scheduling, and control production processes using TESS Project Engineer. She was an industrial relationship coordinator for Industrial Herseg, Torreón, Coahuila (January 1998-June 1998).

Raphael C. W. Phan received the BEng (Honors) from Multimedia University (MMU), Malaysia (1999) and an MEngSc degree by research in "Cryptanalysis of the Advanced Encryption Standard & Skipjack" (2001). He is director of the Information Security Research (iSECURES) Lab with Swinburne University of Technology (Sarawak Campus), Kuching, Malaysia. He reviews for *IEEE Transactions on Computers*, *IEEE Communications Letters*, *IEEE Signal Processing Letters*, *IEEE Transactions on Image Processing*, *IEEE Transactions on Image Processing*, *IEEE Transactions on Wireless Communications*, *Information Processing Letters*, and *Cryptologia*. His research interests include cryptography, cryptanalysis, authentication and key-exchange protocols, side-channel attacks, digital watermarking, and smart card security.

Violina Ratcheva is currently a lecturer in strategic management and enterprise at the University of Sheffield, UK. She has previously worked as an accountant, a lecturer in entrepreneurship, a researcher in European business, and a research fellow in growing SMEs. Her main research interests are in the areas of knowledge creating interactions in interorganizational business networks. She is currently investigating interaction processes and behavior changes in virtual teams, knowledge capturing, and sharing practices in a dispersed organizational environment.

Gergely Sipos is a full-time PhD student at the Jozsef Hatvany PhD School of Information Sciences. Since he received his MSc from the University of Miskolc in 2003, he has worked as a part-time research fellow for the Laboratory of the Parallel and Distributed Systems in the Computer and Automation Research Institute, Hungarian Academy of Sciences. He spent one semester as an MSc student and one semester as a PhD student at the University of Westminster, London. He has published several scientific papers on service-oriented grid environments and tools. He has participated in different Hungarian and international grid projects in the last three years. In 2004 he was awarded with the "best paper" prize by the organizing committee of the MIPRO 2004 Conference.

Xinjian Xu obtained his master of applied science (information technology) degree at the Royal Melbourne Institute of Technology, Australia.

László Zsolt Varga is a senior scientific associate at MTA SZTAKI (Budapest, Hungary). He started research work in the 1980s in KFKI-MSZKI in the field of parallel computing. He was a visiting researcher in the early 1990s in CERN (Geneva) and in the Department of Electronic Engineering at Queen Mary & Westfield College (University of London), where his research focused on basic and applied research into the development of multiagent systems. Currently, he heads a unit in MTA SZTAKI researching and developing distributed knowledge-based applications. His research interests include agent-based and knowledge-based computing for scientific and business applications.

Marcela Velandia worked as a researcher at the Integrated Manufacturing System Research Center (CSIM) of ITESM University, Monterrey, Mexico (January 2002-August 2003). From this institute she also received her MSc degree in manufacturing systems in December 2004. As part of the research center she participated in various projects related with build-to-order supply chains and virtual industry brokerage. Currently, she works for DaimlerChrysler in the Commercial Vehicles Division, Mexico, as an OEM purchaser, where her responsibilities include supplier development and evaluation, supply chain integration, and cost-reduction projects. She has designed and implemented a model for the worldwide supplier evaluation program in the DaimlerChrysler Commercial Vehicles Division.

T. T. Wong received his PhD from the University of Leicester, UK. He is the course manager of the Master of Aviation Program (Massey University) at The Hong Kong Polytechnic University. He is a professional assessment assessor for the Hong Kong Institution of Engineers and the Royal Aeronautical Society. His research interests include information technology and quality management. He is a member of the editorial board of the *International Journal of Industrial Engineering—Theory, Applications and Practice*. He is an expert member of the Six Sigma Promotion Council, China Association for Quality.

Index

A

acquaintance model 62
active
  attackers 339
  knowledge sharing 19
ad hoc networks 169
adhocracy 5
agent 41
  applications 62
  metaphor 41
  model 61
  organizations 53
  reasoning 49
  -oriented software engineering 60
agents with state 47
agreement 55
alignment 117
  and Empowerment 104
AMIS (automated methods for integrating systems) P 245
applicability to smart organizations 282
architecture of the service-oriented grid 295
argumentation 56
artificial intelligence (AI) 43
assistant agents 62, 63
authentication 234, 334
autonomously act 42

B

"Backdoor" attack 240
"Bluebug" attack 240
"Bluejacking" 240
Bluetooth 214
bureaucratic 15
business architect 22

C

chat 337
cluster organization 5
collaboration tools 86
combined service composition 277
COMMA 22
commitment 51
communication 57
  and interaction between partners 167

behaviours 148
in smart organizations 336
networks for smart organizations 198
process 338
communities of practice 124
compatibility with current XML-based standards 283
competence of partners 166
complexity 8
component-based distributed system (CBDS) 268
composing and recomposing Web services 267
composite service layer 276
computer
security institute (CSI) 221
system 42
-mediated communication (CMC) 137
confidentiality 223, 334
continual learning 104, 123
contracting 84
controlling 84
coordination tools 86
CORBA 22
core competence management model 75
COSME 69
cost factor 197
creating alternatives 104
creativity 18
cultural differences 148
customer relationship management systems 112
customization 86

D

DARPA Agent Markup Language (DAML) 23
data conversion (DC) 175
dealing with semantic differences 268
deductive reasoning agents 49
deliberation 51
demand of mobility 191
demands of communication in SO 192
description File 274
desires 51
digital
age 2
economy 188
signatures 344
disciplined decision making 104, 114
distributed agent system 63
dynamism 268

E

e-bidding 84
e-brokerage 82
e-college 22
e-engineering 85
e-marketing 81
e-planning 84
e-productivity 88
e-RFQ (request for quotation) 84
e-selection 83
e-services 78
implementation srchitecture 89
e-supply 87
eavesdropping 169
ECOLNET 22
economic 108
ecosystem of smart organizations 11
efficiency 55
electronic mail (e-mail) 337
empowerment 117
encryption 170, 194, 341
end user requirements 234
ensuring Quality of Service 285
enterprise
architecture 90
information portals 122
environment of implementation 280
essential security technologies 340